40 Topics in Current Chemistry
Fortschritte der chemischen Forschung

Three-Membered Rings

Springer-Verlag
Berlin Heidelberg GmbH 1973

This series presents critical reviews of the present position and future trends in modern chemical research. It is addressed to all research and industrial chemists who wish to keep abreast of advances in their subject.

As a rule, contributions are specially commissioned. The editors aud publishers will, however, always be pleased to receive suggestions and supplementary information. Papers are accepted for ''Topics in Current Chemistry'' in either German or English.

Any volume of the series may be purchased separately.

ISBN 978-3-662-15875-3   ISBN 978-3-540-40013-4 (eBook)
DOI 10.1007/978-3-540-40013-4

Library of Congress Catalog Card Number 51-5497.

Contents

Structures and Stabilities of Three-Membered Rings from *ab initio* Molecular Orbital Theory

Dr. William A. Lathan,* Dr. Leo Radom,** Dr. P. C. Hariharan,
Dr. Warren J. Hehre*** and Prof. Dr. John A. Pople

Department of Chemistry, Carnegie-Mellon University,
Pittsburgh, Pennsylvania, USA

Contents

---

* Current address: Department of Chemistry, University of Rochester, Rochester, New York, USA.

** Current address: Research School of Chemistry, Australian National University, Canberra, Australia.

*** Current address: Laboratoire de Chimie Theorique, Universite de Paris Sud, Orsay, France (Part of Laboratoire de Physico-Chemie des Rayonnements, associated with CNRS).

I. Introduction

The chemistry of three-membered ring molecules is a subject of considerable current interest. Many of the molecules in this class that have been characterized experimentally are only recent discoveries while a large fraction of the possible structures have yet to be synthesized. For this reason, these molecules provide a tempting target for theoretical examination. There have been a number of previous theoretical studies on this subject ranging from early work on the "bent bonding" in cyclopropane by Coulson and Moffitt [1] and by Walsh [2] to recent *ab initio* molecular orbital calculations on several members of the set. [3] In this article, we present results of *ab initio* molecular orbital calculations on the complete set of three-membered ring molecules containing C, N, O and H and which can be written as classical valence structures without charges or unpaired electrons. We have also examined several molecules which are best represented as carbene structures. We attempt to find the lowest singlet states of these molecules. Our main emphasis will be on structures and stabilities although other molecular properties will be discussed in some cases.

II. Method

We use standard *ab initio* self-consistent field (SCF) linear combination of atomic orbitals (LCAO) molecular orbital (MO) theory in this work. The molecular orbitals (ψ_i) are written as linear combinations of basis functions (ϕ_μ)

$$\psi_i = \sum c_{\mu i}\, \phi_\mu \tag{1}$$

The full molecular wave function (Ψ) is then written as a single determinant of the spatial orbitals ψ_i with appropriate spin functions α or β. We will only be concerned with closed shell systems here and for $2n$ electrons in doubly occupied MO's, the wave function is

$$\Psi = [(2n)!]^{-\frac{1}{2}} \, |\psi_1(1)\ \alpha(1)\ \psi_1(2)\ \beta(2) \ldots \psi_n(2n{-}1)\ \alpha(2n{-}1)\ \psi_n(2n)\ \beta(2n)| \tag{2}$$

The total energy may be calculated as

$$E = \int \Psi^* H\, \Psi\, d\tau \tag{3}$$

where H is the many-electron Hamiltonian. Minimization of the energy leads to the Roothaan self-consistent equations [4] for the LCAO coefficients $c_{\mu i}$. The wave function Ψ obtained in this manner then yields the self-consistent value of the total energy from (3). This procedure leads to a molecular energy for any nuclear arrangement once the basis functions ϕ_μ have been specified. The choice of the ϕ_μ is important and is the source of the differences between various *ab initio* molecular orbital calculations.

We use three different basis sets (ϕ_μ) in this work. The first is the minimal STO-3G basis [5] consisting of a least squares fit of three gaussian functions to each function in a minimal basis set ($1s$ on H; $1s$, $2s$, $2px$, $2py$, $2pz$ on C, N, O) of Slater-type orbitals. This basis has been successfully used to calculate geometries of a large number of acyclic molecules [6-10] and some cyclic hydrocarbons.[8] We therefore use it here to obtain optimized structures for each of the three-membered ring molecules.

The second basis set is the split-valence (or extended) 4-31G basis.[11] In this basis set, inner shell orbitals are written as the sum of four gaussian functions while valence orbitals are split into inner and outer parts consisting of three gaussians and one gaussian, respectively. Because the ratio of the inner and outer contributions is free to be determined by the SCF procedure, this basis set provides a more flexible description of the electronic distribution than STO-3G. It has proved more reliable in energy comparisons than STO-3G.[8,12-14] We therefore carry out for this purpose, single 4-31G calculations at the STO-3G optimized geometries for each molecule.

In comparing energies of acyclic and cyclic molecules, it is found that even the 4-31G basis does not always give adequate results.[8,12] Indeed, acyclic molecules are predicted to be relatively too stable with the 4-31G basis set. Recent calculations [15] have shown that addition of polarization functions (d-type functions) to the basis functions for heavy (non-hydrogen) atoms leads to improved results in energy comparisons of this type. We have therefore carried out, in a number of cases, single calculations at the STO-3G optimized geometries with such a basis. This basis set [16] is denoted 6-31G*. It consists of 6-31G basis functions [17] (analogous to 4-31G described above) together with d_{x2}, d_{y2}, d_{z2}, d_{xy}, d_{yz} and d_{xz} functions for the heavy atoms. These six second degree gaussians correspond to the five pure d-type gaussians (xy, xz, yz, $3z^2-r^2$, x^2-y^2)$e^{-a_d r^2}$ together with the $3s$ type gaussian ($x^2+y^2+z^2$)$e^{-a_d r^2}$. The polarization exponent α_d is taken as 0.8 for all three heavy atoms (C, N, O) on the basis of optimization studies.[16]

In order to obtain our theoretical geometries, we begin by specifying a symmetry that defines the structure in which we are interested. Each independent geometric parameter is then varied in turn so as to minimize the total energy. This process is continued until the energy is a minimum with respect to all these geometric parameters. Expectation values of the electric dipole moment of each molecule are calculated by standard procedures. Electron populations are calculated using Mulliken's method.[18]

III. Results

Table 1 presents calculated energies for the singlet states of the set of three-membered rings. In addition to total energies and relative energies, we have also listed bond separation energies [12] for some molecules. These are the energy changes in the formal reactions in which the three-membered rings are converted into molecules with two heavy atoms and the same types of bonds. For example, the bond separation reaction for cyclopropene is

$$
\underset{HC=\!=\!CH}{\overset{\overset{\displaystyle H_2}{\overset{\displaystyle C}{\diagup\ \diagdown}}}{}} + 3\,CH_4 \longrightarrow H_2C\!=\!CH_2 + 2\,H_3C\!-\!CH_3
$$

The bond separation energies reflect the stabilization or destabilization that results when the separated bonds between heavy atoms are brought together in the three-membered ring.

For the three-membered ring molecules, the bond separation energies depend primarily on three factors:

(1) The strain energy of the ring,

(2) the stabilizing or destabilizing interaction between the adjacent bonds, and

(3) the stability or instability associated with the π-electron structure of the ring.

In some cases (*e.g.* cyclopropane), one of these factors (*e.g.* ring strain) is likely to be much more important than the other two. In other cases, all three factors may have a significant influence on the bond separation energy.

Table 2 gives the electric dipole moments calculated with the three basis sets. Table 3 gives the 6-31G* calculated orbital energies. These may prove useful in interpreting the photoelectron spectra of these molecules.[19] Finally, the optimized geometries themselves, as obtained with the STO-3G basis set, are presented within the text of this paper as structures *1—32*.

5

Table 1. Calculated energy data for three-membered rings

Stoichiometric formula	Name	Structural formula	STO-3G Total energy (hartrees)	4-31G Total energy (hartrees)	Relative energy[1] (kcal/mol)	Bond separation energy (kcal/mol)	6-31G* Total energy (hartrees)	Relative energy[1] (kcal/mol)	Bond separation energy (kcal/mol)
C_3H_6	Cyclopropane (1)	$\begin{array}{c} H_2 \\ C \\ H_2C\!-\!CH_2 \end{array}$	−115.66616	−116.88350		−28.0	−117.05872		−26.2
C_3H_4	Cyclopropene (2)	$\begin{array}{c} H_2 \\ C \\ HC\!=\!CH \end{array}$	−114.40116	−115.64168	0	−58.1	−115.82294	0	−50.4
C_3H_4	Cyclopropylidene (3)	$\begin{array}{c} \ddot{C} \\ H_2C\!-\!CH_2 \end{array}$	−114.35525	−115.58345	36.5		−115.76585	35.8	
C_3H_2	Cyclopropenylidene (4)	$\begin{array}{c} H_2 \\ C \\ HC\!=\!CH \end{array}$	−113.17804	−114.42426	0		−114.61849		
C_3H_2	Cyclopropyne (5)	$\begin{array}{c} H_2 \\ C \\ C\!\equiv\!C \end{array}$	−112.99883[2]	−114.28388[2]	88.1	−149.6			
C_3	Cyclopropynylidene (6)	$\begin{array}{c} \ddot{C} \\ C\!\equiv\!C \end{array}$	−111.81969[2]	−113.08167[2]		−90.8			

C_2NH_5	Aziridine (7)	NH / CH_2—CH_2	−131.39947	−132.81976		−30.2	−133.03524		−22.3
C_2NH_3	1-Azirine (8)	CH_2 / CH = N	−130.17761	−131.61075	0	−45.0	−131.83696	0	−33.1
C_2NH_3	Aziridinylidene (9)	NH / :C — CH_2	−130.13203	−131.57796	20.6				
C_2NH_3	2-Azirine (10)	NH / CH = CH	−130.11296	−131.55598	34.4	−74.1	−131.77242	40.5	−63.5
C_2NH	Azirinylidene (11)	C̈ / HN = N	−128.94929	−130.37858	0				
C_2NH	Aziridinediylidene (12)	NH / :C = C:	−128.89236	−130.33612	26.6				
C_2OH_4	Oxirane (13)	O / CH_2 — CH_2	−150.92850	−152.62444		−26.9	−152.86456		−19.2
C_2OH_2	Oxiranylidene (14)	O / CH_2 — C:	−149.67634	−151.36715	0				
C_2OH_2	Oxirene (15)	O / CH = CH	−149.62425	−151.35035	10.5	−77.2	−151.58162		−73.0

Table 1 (continued)

Stoichiometric formula	Name	Structural formula	STO-3G Total energy (hartrees)	4-31G Total energy (hartrees)	4-31G Relative energy[1] (kcal/mol)	4-31G Bond separation energy (kcal/mol)	6-31G* Total energy (hartrees)	6-31G* Relative energy[1] (kcal/mol)	6-31G* Bond separation energy (kcal/mol)
C_2O	Oxiranediylidene (16)	O \ :C — C:	−148.41878	−150.09925					
CN_2H_4	Diaziridine trans (17A)	H_2C \ HN—NH	−147.12903	−148.74266	0	−27.9	−148.99571		−14.5
	cis (17B)		−147.11771	−148.73016	7.8	−35.7			
CN_2H_2	3 H-Diazirine (18)	H_2C \ N=N	−145.94650	−147.55161	0	−29.7	−147.82563	0	−10.7
CN_2H_2	1 H-Diazirine (19)	H–N \ HC=N	−145.89176	−147.52007	19.8	−51.2	−147.78114	27.9	−35.5
CN_2H_2	Diaziridinylidene trans (20A)	:C̈ \ HN—NH	−145.86324	−147.48481	41.9				
	cis (20B)		−145.85404	−147.47262	49.6				
CN_2	Diazirinylidene (21)	:C̈ \ N=N	−144.71367	−146.30862					

Formula	Compound	Structure						
CNOH₃	Oxaziridine (22)	$\mathrm{H_2C}$ ⟍ O, HN—O	−166.64623		−168.52715	−30.3	−168.80321	−15.7
CNOH	Oxazirine (23)	$\mathrm{H\!-\!C}$ ⟍ O, N—O	−165.40629	0	−167.31203	−48.9	−167.58675	−37.9
CNOH	Oxaziridinylidene (24)	$\mathrm{\ddot{C}}$ ⟍ O, HN—O	−165.40406	20.4	−167.27946			
CO₂H₂	Dioxirane (25)	$\mathrm{H_2C}$ ⟍ O, O—O	−186.16142		−188.30285	−23.4	−188.59733	−10.7
CO₂	Dioxiranylidene (26)	$\mathrm{\ddot{C}}$ ⟍ O, O—O	−184.93281		−187.06714			
N₃H₃	Triaziridine trans (27A)	$\mathrm{H\!-\!N}$ ⟍ NH, HN—NH	−162.84555	0	−164.64027	−28.5	−164.93092	−8.3
	cis (27B)		−162.82214	17.0	−164.61311	−45.5		
N₃H	Triazirine (28)	$\mathrm{H\!-\!N}$ ⟍ N, N=N	−161.66348		−163.45196	−28.5	−163.76178	−3.9

Table 1 (continued)

Stoichio-metric formula	Name	Structural formula	STO-3G Total energy (hartrees)	4-31G Total energy (hartrees)	Relative energy[1] (kcal/mol)	Bond separation energy (kcal/mol)	6-31G* Total energy (hartrees)	Relative energy[1] (kcal/mol)	Bond separation energy (kcal/mol)
N$_2$OH$_2$	Oxadiaziridine *trans* (29A)		-182.36201	-184.41407		-30.6	-184.72404		-9.3
	cis (29B)		-182.35053	-184.40001		-39.5			
N$_2$O	Oxadiazirine (30)		-181.18013	-183.23790		-23.1	-183.59930		-2.2
NO$_2$H	Dioxaziridine (31)		-201.86561	-204.16300		-33.5	-204.49239		-11.2
O$_3$	Trioxirane (32)		-221.36764	-223.89518		-32.0	-224.24260		-9.7

[1] Energy relative to that of the most stable structural isomer.

[2] The wave function is found to be complex.

Table 2. Calculated dipole moments (debyes) for three-membered rings

Molecule	STO-3G	4-31G	6-31G*
Cyclopropane	0	0	0
Cyclopropene	0.55	0.56	0.57
Cyclopropylidene	1.43	2.14	2.13
Cyclopropenylidene	2.62	3.34	3.35
Cyclopropyne	2.33	2.94	
Cyclopropynylidene	0.70	0.98	
Aziridine	1.82	2.45	2.09
1-Azirine	1.92	2.77	2.56
Aziridinylidene	2.39	2.82	
2-Azirine	2.11	2.85	2.51
Azirinylidene	2.42	3.23	
Aziridinediylidene	2.83	3.20	
Oxirane	1.46	2.90	2.43
Oxiranylidene	1.42	2.53	
Oxirene	1.72	3.31	2.96
Oxiranediylidene	0.40	0.43	
Diaziridine (trans)	1.33	1.99	1.72
Diaziridine (cis)	3.20	4.15	
3 H-Diazirine	1.58	2.44	2.11
1 H-Diazirine	2.78	3.82	3.50
Diaziridinylidene (trans)	0.59	0.62	
Diaziridinylidene (cis)	2.85	3.64	
Diazirinylidine	0.12	0.14	
Oxaziridine	2.19	3.54	3.03
Oxazirine	2.68	4.23	3.93
Oxaziridinylidene	1.47	2.24	
Dioxirane	1.93	3.73	3.23
Dioxiranylidene	2.28	1.30	
Triaziridine (trans)	1.47	1.87	1.59
Triaziridine (cis)	4.45	5.63	
Triazirine	1.51	2.04	1.71
Oxadiaziridine (trans)	0.50	1.57	1.35
Oxadiaziridine (cis)	3.06	4.22	
Oxadiazirine	1.76	1.19	1.13
Dioxaziridine	1.66	2.73	2.33
Trioxirane	0	0	0

11

Table 3. Calculated (6-31G*) orbital energies (hartrees)[1]

Cyclopropane (D_{3h})	Cyclopropene (C_{2v})	Cyclopropylidene (C_{2v})
-11.2216 ($1a_1$)	-11.2417 ($1a_1$)	-11.2874 ($1a_1$)
-11.2205 ($1e'$)	-11.2393 ($1b_1$)	-11.2391 ($2a_1$)
-11.2205 ($1e'$)	-11.2340 ($2a_1$)	-11.2384 ($1b_1$)
-1.1302 ($2a_1'$)	-1.1630 ($3a_1$)	-1.1430 ($3a_1$)
-0.8163 ($2e'$)	-0.8216 ($4a_1$)	-0.8389 ($2b_1$)
-0.8163 ($2e'$)	-0.7635 ($2b_1$)	-0.7454 ($4a_1$)
-0.6649 ($1a_2''$)	-0.6796 ($5a_1$)	-0.6523 ($1b_2$)
-0.6271 ($3a_1'$)	-0.5933 ($1b_2$)	-0.5562 ($5a_1$)
-0.5097 ($1e''$)	-0.4929 ($6a_1$)	-0.5302 ($1a_2$)
-0.5097 ($1e''$)	-0.4171 ($3b_1$)	-0.4252 ($3b_1$)
-0.4169 ($3e'$)	-0.3551 ($2b_2$)	-0.3595 ($6a_1$)
-0.4169 ($3e'$)		

Cyclopropenylidene (C_{2v})	Aziridine (C_s)	1-Azirine (C_s)
-11.2815 ($1a_1$)	-15.5644 ($1a'$)	-15.5977 ($1a'$)
-11.2796 ($1b_1$)	-11.2563 ($2a'$)	-11.3009 ($2a'$)
-11.2508 ($2a_1$)	-11.2556 ($1a''$)	-11.2722 ($3a'$)
-1.2038 ($3a_1$)	-1.2481 ($3a'$)	-1.3011 ($4a'$)
-0.7975 ($2b_1$)	-0.8892 ($4a'$)	-0.8747 ($5a'$)
-0.7753 ($4a_1$)	-0.8388 ($2a''$)	-0.8042 ($6a'$)
-0.6201 ($5a_1$)	-0.6926 ($5a'$)	-0.6451 ($7a'$)
-0.4748 ($1b_2$)	-0.6311 ($6a'$)	-0.6318 ($1a''$)
-0.4578 ($3b_1$)	-0.5284 ($3a''$)	-0.5339 ($8a'$)
-0.3679 ($6a_1$)	-0.4762 ($7a'$)	-0.4146 ($2a''$)
	-0.4746 ($4a''$)	-0.4050 ($9a'$)
	-0.3898 ($8a'$)	

2-Azirine (C_s)	Oxirane (C_{2v})	Oxirene (C_{2v})
-15.5740 ($1a'$)	-20.5770 ($1a_1$)	-20.5713 ($1a_1$)
-11.2765 ($2a'$)	-11.2885 ($2a_1$)	-11.3066 ($2a_1$)
-11.2740 ($1a''$)	-11.2877 ($1b_1$)	-11.3040 ($1b_1$)
-1.2533 ($3a'$)	-1.4048 ($3a_1$)	-1.3750 ($3a_1$)
-0.9041 ($4a'$)	-0.9304 ($4a_1$)	-0.9615 ($4a_1$)
-0.7882 ($2a''$)	-0.8614 ($2b_1$)	-0.8103 ($2b_1$)
-0.6996 ($5a'$)	-0.6983 ($1b_2$)	-0.7134 ($5a_1$)
-0.5848 ($6a'$)	-0.6579 ($5a_1$)	-0.5773 ($1b_2$)
-0.4760 ($7a'$)	-0.5477 ($3b_1$)	-0.5144 ($3b_1$)
-0.4577 ($3a''$)	-0.5424 ($1a_2$)	-0.5066 ($6a_1$)
-0.3285 ($8a'$)	-0.4483 ($2b_2$)	-0.3340 ($2b_2$)
	-0.4425 ($6a_1$)	

Table 3 (continued)

trans-Diaziridine (C_2)	3 H-Diazirine (C_{2v})	1 H-Diazirine (C_1)
−15.6013 (1a)	−15.6806 (1a₁)	−15.6332 (1a)
−15.6009 (1b)	−15.6793 (1b₁)	−15.6194 (2a)
−11.2964 (2a)	−11.3130 (2a₁)	−11.3465 (3a)
−1.3363 (3a)	−1.4473 (3a₁)	−1.3678 (4a)
−0.9751 (2b)	−0.9172 (4a₁)	−0.9979 (5a)
−0.8812 (4a)	−0.8552 (2b₁)	−0.8347 (6a)
−0.6896 (3b)	−0.6778 (1b₂)	−0.6750 (7a)
−0.6621 (5a)	−0.6557 (5a₁)	−0.6096 (8a)
−0.5361 (4b)	−0.5714 (6a₁)	−0.5284 (9a)
−0.5123 (6a)	−0.4763 (2b₂)	−0.4489 (10a)
−0.4312 (5b)	−0.4177 (3b₁)	−0.3903 (11a)
−0.4134 (7a)		

Oxaziridine (C_1)	Oxazirine (C_s)	Dioxirane (C_{2v})
−20.6150 (1a)	−20.6210 (1a′)	−20.6537 (1a₁)
−15.6389 (2a)	−15.6669 (2a′)	−20.6533 (1b₁)
−11.3348 (3a)	−11.3894 (3a′)	−11.3769 (2a₁)
−1.4670 (4a)	−1.4672 (4a′)	−1.5644 (3a₁)
−1.0448 (5a)	−1.0952 (5a′)	−1.1569 (2b₁)
−0.9095 (6a)	−0.8650 (6a′)	−0.9395 (4a₁)
−0.7292 (7a)	−0.6728 (7a′)	−0.7428 (1b₂)
−0.6642 (8a)	−0.6311 (1a″)	−0.6982 (5a₁)
−0.5728 (9a)	−0.5576 (8a′)	−0.5898 (6a₁)
−0.5172 (10a)	−0.5084 (9a′)	−0.5437 (2b₂)
−0.4895 (11a)	−0.4039 (2a″)	−0.5407 (3b₁)
−0.4311 (12a)		−0.4621 (1a₂)

trans-Triaziridine (C_s)	Triazirine (C_s)	trans-Oxadiaziridine (C_2)
−15.6476 (1a′)	−15.7334 (1a′)	−20.6592 (1a)
−15.6471 (1a″)	−15.7320 (1a″)	−15.6901 (2a)
−15.6462 (2a′)	−15.6674 (2a′)	−15.6897 (1b)
−1.4169 (3a′)	−1.5085 (3a′)	−15.3320 (3a)
−1.0024 (4a′)	−1.0629 (4a′)	−1.0914 (4a)
−0.9984 (2a″)	−0.8794 (2a″)	−1.0298 (2b)
−0.7224 (5a′)	−0.7064 (5a′)	−0.7373 (3b)
−0.6617 (6a′)	−0.6338 (6a′)	−0.6937 (5a)
−0.5510 (3a″)	−0.5958 (7a′)	−0.5734 (4b)
−0.4994 (7a′)	−0.4664 (3a″)	−0.5451 (6a)
−0.4480 (8a′)	−0.4550 (8a′)	−0.4696 (7a)
−0.4316 (4a″)		−0.4671 (5b)

13

Table 3 (continued)

Oxadiazirine (C_{2v})	Dioxaziridine (C_s)	Trioxirane (D_{3h})
−20.6786 ($1a_1'$)	−20.7081 ($1a'$)	−20.7689 ($1a_1'$)
−15.7797 ($2a_1$)	−20.7077 ($1a''$)	−20.7683 ($1e'$)
−15.7781 ($1b_1$)	−15.7427 ($2a'$)	−20.7683 ($1e'$)
−1.5886 ($3a_1$)	−1.6328 ($3a'$)	−1.7312 ($2a_1'$)
−1.1974 ($4a_1$)	−1.1900 ($2a''$)	−1.2331 ($2e'$)
−0.9090 ($2b_1$)	−1.0898 ($4a'$)	−1.2331 ($2e'$)
−0.6964 ($1b_2$)	−0.7826 ($5a'$)	−0.8132 ($1a_2''$)
−0.6764 ($5a_1$)	−0.6997 ($6a'$)	−0.7447 ($3a_1'$)
−0.6299 ($6a_1$)	−0.6237 ($7a'$)	−0.6524 ($3e'$)
−0.5384 ($3b_1$)	−0.5843 ($3a''$)	−0.6524 ($3e'$)
−0.4790 ($2b_2$)	−0.5042 ($8a'$)	−0.5469 ($1e''$)
	−0.5029 ($4a''$)	−0.5469 ($1e''$)

[1]) For the C_{2v} group, the symmetry species a_2 and b_2 are antisymmetric with respect to the plane of the ring.

IV. Comparative Data for Acyclic Systems

In order to calculate bond separation energies for the three-membered rings, we need the total energies for certain molecules with one and two heavy atoms calculated using comparable procedures. These are summarized in Table 4. Also given is a comparison of STO-3G optimized and experimental bond lengths for the one and two heavy atom systems. Finally, we present calculated and experimental dipole moments for these systems. The bond length and dipole moment comparisons will be useful in assessing our results for the larger, cyclic molecules.

V. Comparison with Experimental Data

So that we may gain some idea of the value of our predictions for molecules for which experimental information is not available, it is important to compare theoretical and experimental results in the cases where the latter *is* known. We present here such a comparison for geometries and energies.

1. Geometries

Well characterized experimental structures are currently available for only five of the three-membered rings, cyclopropane [20] (*1*), cyclo-

Table 4. Comparative data for acyclic molecules

Molecule	Bond	Bondlength (Å)[1]		Total energy (hartrees)		ΔH_f° (298 °K)[2] expt (kcal/mol)	Dipole moment (debyes)			
		Calc.	Expt.	4-31G[1]	6-31G*		STO-3G[1]	4-31G[1]	6-31G*	Expt.[3]
:CH₂ (¹A₁)	C—H	1.123	1.11	−38.80932	−38.87152		1.54	2.15	1.97	
CH₄	C—H	1.083	1.085	−40.13976	−40.19517	−17.889	0	0	0	0
NH₃	N—H	1.033	1.012	−56.09829	−56.18190	−11.02	1.88	2.44	2.08	1.47
OH₂	O—H	0.990	0.957	−75.90324	−76.00678	−57.796	1.71	2.69	2.33	1.85
HC≡CH	C≡C	1.168	1.203	−76.70999		54.194	0	0	0	0
H₃C—ĊH (¹A')	C—C	1.537		−77.80265	−77.92030		1.72	2.35	2.19	
H₂C=CH₂	C=C	1.306	1.330	−77.92188	−78.03145	12.496	0	0	0	0
H₃C—CH₃	C—C	1.538	1.531	−79.11582	−79.22865	−20.236	0	0	0	0
HĊ—NH₂ (¹A')	:C—N	1.339		−93.83283			2.70	3.26		
H₂C=NH	C=N	1.273		−93.87947	−94.02611		1.85	2.56	2.33	
H₃C—NH₂	C—N	1.486	1.474	−95.06498	−95.20722	−5.49	1.62	2.06	1.74	1.31
HĊ—OH (¹A')	:C—O	1.331		−113.60763			1.21	1.00		
H₃C—OH	C—O	1.433	1.427	−114.86716	−115.03180	−47.96	1.51	2.45	2.04	1.70
HN=NH	N=N	1.267	1.238	−109.80525	−109.98717	50.2	0	0	0	0
H₂N—NH₂	N—N	1.459	1.453	−110.99350	−111.16330	22.80	2.22	2.82	2.43	1.75
H₂N—OH	N—O	1.420	1.46	−130.78462	−130.97310	−9.3	0.41	0.87	0.73	
HO—OH	O—O	1.396	1.475	−150.55198	−150.75944	−32.58	1.30	2.06	1.75	2.26

[1] For one heavy atom systems, taken from Ref. [9]; for two heavy atom systems, taken from Ref. [10]. All values refer to the singlet state structure of lowest energy for the particular molecule.

[2] Summarized in Ref. [13].

[3] Nelson, R. D., Lide, D. R., Maryott, A. A.: Selected values of electric dipole moments for molecules in the gas phase, NSRDS-NBS 10. Washington, D. C.: U.S. Government Printing Office 1967.

propene [21] (*2*), aziridine [22] (*7*), oxirane [23] (*13*) and 3H-diazirine [24] (*18*). The theoretical and experimental geometries for these molecules are compared in Fig. 1. General agreement is good. Discrepancies that do

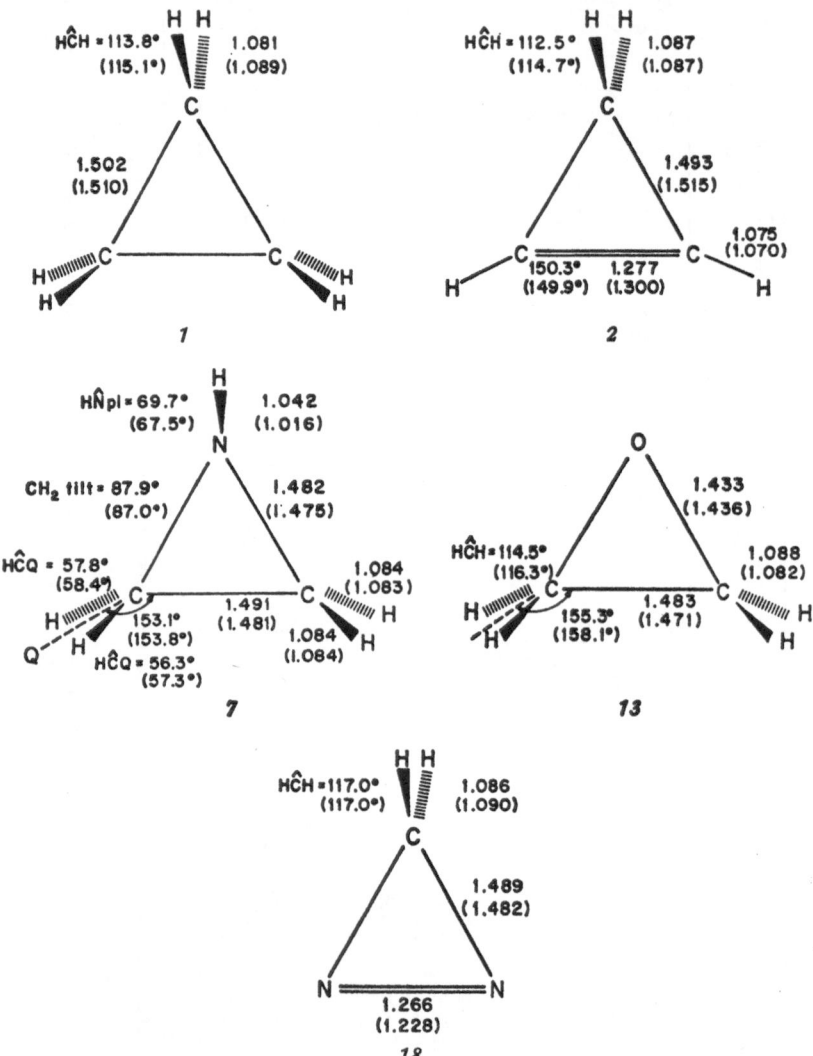

Fig. 1. Comparison of calculated and experimental (in parentheses) geometries (Throughout this paper, all bond lengths are in Å, the notation *HNpl* is used to denote the angle between an N—H bond and the ring plane, the line *CQ* represents the intersection of HCH and ring planes and *CH₂ tilt* denotes the angle between the HCH plane and the plane of the ring)

occur closely parallel corresponding results for one and two heavy atom systems (Table 4). Thus, whereas C—H bonds are in close agreement with experiment, N—H lengths are too long. The predicted C=C double bond length in cyclopropene is too short, the difference being similar to that found for ethylene. Similarly, the length of the double bond in 3 H-diazirine is overestimated as is the N=N double bond in diimide. The lengths of single bonds between heavy atoms agree well with experimental values except for the C—C bond in cyclopropene which is too short. Angles are reproduced well, even the methylene tilt in aziridine. For the nine unique bonds between heavy atoms, the mean absolute deviation between theoretical and experimental bond lengths is only 0.014 Å.

2. Energies

Experimental heats of formation are available for five of the three-membered rings; calculated and experimental bond separation energies for these molecules are compared in Table 5. Three sets of calculated values are listed corresponding to the STO-3G, 4-31G and 6-31G* basis sets. The theoretical bond separation energies should be compared with experimental values at 0 °K corrected for zero-point vibration; these are quoted in the last column of the table. However, since the data (*e.g.* vibrational frequencies) required to obtain such experimental values are not always available, we also give experimental bond separation energies at 0 °K and 298 °K without the vibrational corrections in order to point out the approximate magnitude of such corrections.

Table 5. Calculated and experimental bond separation energies (kcal/mol)

Molecule	Calculated			Experimental[1]		
	STO-3G	4-31G	6-31G*	298°	0 °K	0 °K with vibrational correction
Cyclopropane	−45.1	−28.0	−26.2	−19.8	−18.5	−23.5
Cyclopropene	−65.6	−58.1	−50.4	−40.9	−39.3	−45.2
Aziridine	−39.7	−30.3	−22.3	−14.6	−13.2	−19.3
Oxirane	−35.4	−26.9	−19.2	−10.0	− 8.6	−14.0
3 H-Diazirine	−24.1	−29.7	−10.7	+ 0.1		

[1] Calculated using experimental ΔH_f° (298 °K) values for three-membered rings from Table 6 and vibrational frequencies from Eggers, D. F., Schultz, J. W., Wiberg, K. B., Wagner, E. L., Jackman, L. M., Erskine, R. L.: J. Chem. Phys. *47*, 946 (1967) (cyclopropene), and Shimanouchi, T.: Tables of molecular vibrational frequencies, NSRDS-NBS 11. Washington, D. C.: U.S. Government Printing Office 1967 (remaining cyclic molecules). Experimental data for one and two heavy atom molecules taken from Ref. [12]. Method of applying vibrational corrections summarized in Ref. [12].

The theoretical results are always too negative, *i.e.* the cyclic molecules are predicted to be too unstable relative to the two heavy atom products of the bond separation reaction. However, there is a steady improvement in going from STO-3G (mean deviation = 21 kcal/mol to 4-31G (mean deviation = 10 kcal/mol) to 6-31G* (mean deviation = 4 kcal/mol). We also note (for use in Section VI) that the experimental bond separation energies at 298 °K are always less negative than the vibrationally corrected values at 0 °K; the mean difference between these sets of values is 4.2 ± 0.5 kcal/mol.

VI. Calculation of Heats of Formation

If ΔH_f° (298 °K) values for all the species with one or two heavy atoms in the bond separation reaction are known *and* if the bond separation energy at 298 °K can be predicted from theoretical calculations, the heat of formation of the three-membered ring can be estimated. In Table 6, we present heats of formation calculated in this manner from experimental heats of formation for molecules with one and two heavy atoms from Table 4 and our theoretical bond separation energies from Table 1. The

Table 6. Calculated and experimental heats of formation (ΔH_f° (298 °K), kcal/mol)

Molecule	6-31G*	6-31G*[1]) (corr.)	Expt.
Cyclopropane (1)	19.2	11	12.7[2])
Cyclopropene (2)	76.1	68	66.6[2])
Aziridine (7)	37.9	30	30.2[3])
2-Azirine (10)	111.8	104	
Oxirane (13)	−3.4	−12	−12.6[3])
Oxirene (15)	83.2	75	
Diaziridine (trans) (17)	66.2	58	
3 H-Diazirine (18)	89.8	82	79.3[2])
Oxaziridine (22)	39.7	32	
Dioxirane (25)	15.7	8	
Triaziridine (trans) (27)	109.8	102	
Triazirine (28)	132.8	125	
Oxadiaziridine (trans) (29)	93.3	85	
Oxadiazirine (30)	113.6	105	
Dioxaziridine (31)	86.6	78	
Trioxirane (32)	85.3	77	

[1]) Empirically corrected values (see Section VI of text).
[2]) From Ref. [26].
[3]) From Ref. [25].

first column lists heats of formation calculated using 6-31G* bond separation energies assuming that these values are applicable at 298 °K. However, as we noted in Section V.2, the calculated bond separation energies refer to heats of reaction at 0 °K with stationary nuclei. We found, in the first place, a systematic difference $(4.2 \pm 0.5$ kcal/mol) between experimental bond separation energies under these conditions and values at 298 °K. In addition, we found that the calculated bond separation energies, even at the 6-31G* level were uniformly too negative by 4.0 ± 1.3 kcal/mol. Although there is not really sufficient experimental data to make a very reliable assessment of the magnitude of these errors, we do believe they are systematic rather than random. We therefore include in Table 6, an additional set of theoretical heats of formation corresponding to the original 6-31G* values empirically corrected by 8.2 kcal/mol. We consider that these are better than the uncorrected estimates of the heats of formation.

VII. Discussion of Individual Three-Membered Ring Molecules

In the previous sections, we have compared our theoretical results with experiment in the few cases where experimental information is available. This provides the background now for a discussion of the individual molecules, many of which are not well-characterized experimentally.

1. C_3H_6 Cyclopropane (1)

Cyclopropane is known to have a symmetrical D_{3h} structure with the three carbon atoms at the vertices of an equilateral triangle. Our results on the STO-3G geometry, which have already been published [8] are in close agreement with the experimental structure (see Fig. 1). The C—C lengths are found to be shorter than in acyclic alkanes and the HCH angles to be opened up beyond the tetrahedral value. There have been a large number of previous studies of cyclopropane using *ab initio* molecular orbital theory.[12,27-42] The geometry studies that have been reported have given results similar to ours. Preuss and Diercksen,[27] using a smaller gaussian basis found 1.54 Å for the C—C length in a partial geometry optimization. Frost and Rouse [28] made a complete study (assuming D_{3h} symmetry) using floating spherical gaussian orbitals and found a structure similar to ours. Buenker and Peyerimhoff [29] varied one CCC angle to reduce the symmetry to C_{2v} (assuming values for other parameters). They found an apparent minimum corresponding to a CCC angle of 63.5°, deviating slightly from the generally assumed D_{3h} symmetry. However, more detailed calculations with the STO-3G basis yield no evidence of such a distortion.[8]

It is generally considered that cyclopropane, along with other three-membered rings, is destabilized by ring strain. The (negative) energy of the bond separation reaction

$$\overline{CH_2 \cdot CH_2 \cdot CH_2} + 3\,CH_4 \longrightarrow 3\,CH_3 \cdot CH_3$$

which compares the energy of C—C bonds in cyclopropane with separated C—C bonds as in ethane is largely a measure of the ring strain although other factors such as the stabilizing effect of adjacent C—C bonds undoubtedly contribute. Experimental values of the bond separation energy for cyclopropane are -18.5 kcal/mol at $0\,°K$ or -23.5 kcal/mol after correction for zero-point vibrations. The calculated bond separation energies approach this value as the basis set is improved (STO-3G $\rightarrow$ 4-31G $\rightarrow$ 6-31G*, see Table 5) but there remains a small residual error.

The energy of cyclopropane may also be compared with that of its more stable open-chain isomer, propene. The relative energy is 7.3 kcal/mol calculated using experimental heats of formation at $0\,°K$ (7.4 kcal/mol after zero-point correction). Calculated values of this energy difference are -3.7 kcal/mol (STO-3G), $+13.2$ kcal/mol (4-31G) and $+7.8$ kcal/mol (6-31G*). The STO-3G result is clearly unsatisfactory. It seems to be characteristic of a general deficiency of the isotropic minimal basis which gives a better description of single than multiple bonds. According to the 4-31G extended sp basis, the difference is positive but is too large. The same is found using the 6-31G basis with improved inner shells. However, the addition of d-functions to the basis (6-31G $\rightarrow$ 6-31G*) preferentially lowers the energy of the ring compound and the final energy difference is in good agreement with experiment.[15]

2. C₃H₄ Cyclopropene (2)

Cyclopropene is known to be a symmetrical (C_{2v}) molecule with an isoceles triangular carbon framework. The STO-3G structure (assuming C_{2v} symmetry) has already been published.[8] The calculated length for the C=C bond is too short (see Fig. 1) as is also found for ethylene. However, both theory and experiment show a shortening of about 0.03 Å in going from ethylene to cyclopropene. The theoretical C—C bonds are also somewhat too short but the shape of the CCC triangle is given well (the calculated CCC angle at the apex is 50.7° vs. 50.8° experimentally). Peyerimhoff and Buenker [42] have found a value of 53.4° for this angle assuming values for all other parameters.

The bond separation energy for cyclopropene is considerably more negative than for cyclopropane suggesting increased strain energy. This is presumably associated with the greater distortion of the CCC angles from the values in corresponding acyclic compounds. Again, the theory reproduces the bond separation energy moderately well at the 6-31G* level, [15] but there is a residual error of about 5 kcal/mol. The energies of cyclopropene relative to the open-chain isomers propyne and allene are listed in Table 7. The theoretical results show features similar to

Table 7. Relative energies (kcal/mol) for C_3H_4 isomers

Molecule	PB[1]	C[2]	STO-3G[3]	4-31G[3]	6-31G[3]	6-31G*[4]	Expt.[5]
Propyne	0	0	0	0	0	0	0
Allene	−1.7		17.1	0.8	0.4	1.7	2.1
Cyclopropene	34.0	37.0	30.0	36.4	35.2	25.4	22.3
Cyclopropylidene			58.8	72.9		61.2	

[1] Experimental geometries, Ref. [42].
[2] Experimental geometries, Ref. [30].
[3] STO-3G optimized geometries, Ref. [8].
[4] STO-3G optimized geometries, Ref. [15].
[5] As summarized in Ref. [8].

the C_3H_6 compounds. The sp basis (4-31G) leads to values for the energy of cyclopropene which are relatively too high, but this is largely corrected by addition of d-functions.

A previous study of the charge distribution in cyclopropene using the 6-31G basis set suggested that π-electrons are *withdrawn* from the double bond into the methylene group.[8] This led to a theoretical dipole moment with its negative end at the CH_2 group and the following π-orbital populations

Addition of d-functions with the 6-31G* basis does not alter the sense of this moment. The magnitude of the theoretical dipole with this basis is 0.57 debyes which is in reasonable agreement with the experimental value [21] of 0.45 debyes. Similar results were obtained by other *ab initio* studies [12,30,31,34–36,38,43–46] and some of these are listed in Table 8.

The theoretical polarity of cyclopropene is the reverse of the usual polarity involving the interaction of saturated and unsaturated hydro-

Table 8. Energies and dipole moments for cyclopropene

Authors	Energy (hartrees)	Dipole moment (debyes)	Ref.
Clark	−114.7725	0.48	30)
André *et al.*	−115.2714	0.58	45)
Bonaccorsi *et al.*	−115.4973	0.34	35)
Kochanski and Lehn	−115.7572	0.47	36)
Peyerimhoff and Buenker	−115.7635	—	43)
Robin *et al.*	−115.7655	0.36	44)
This work (6-31G*)	−115.8229	0.57	
Expt.	—	0.45	21)

carbon fragments. Its origin can be understood qualitatively by noting that the empty antibonding π^* orbital associated with the C=C group is of a_2 symmetry (point group C_{2v}) and is therefore not available to accept electrons from the π-like orbitals of the CH_2 group which all have b_2 symmetry. The only electron transfer that can occur (within the valence orbital framework), therefore, is from the occupied bonding π orbital of C=C (symmetry b_2) to the antibonding orbitals of the CH_2 group. This appears to be the main effect giving rise to the theoretical dipole. Benson and Flygare [47] have found the *opposite* dipole direction in cyclopropene from the experimental g values. This discrepancy between theory and experiment is unresolved at the present time.

3. C_3H_4 Cyclopropylidene (*3*)

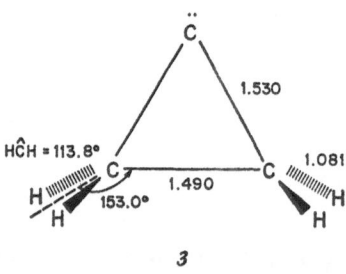

3

This is the other possible cyclic isomer of C_3H_4 and may be considered as a cyclic carbene. Only the singlet state has been considered, although it is possible that an associated triplet state may be lower in energy. C_{2v} symmetry was assumed in the structure determination. The cyclopropylidene skeleton is rather similar to that of cyclopropane, from which cyclopropylidene can be formally derived by abstraction of two geminal hydrogen atoms. There is a slight reduction in the CCC angle at the carbene center.

The energy studies suggest that this isomer is less stable than cyclopropene (by 37 kcal/mol according to the 6-31G* basis). It is 61 kcal/mol less stable than allene, to which it may be transformed by simple cleavage of a C–C bond. The path of such a ring-opening would involve lowering the symmetry from C_{2v} and has not been investigated.[48]

The calculated dipole moment of cyclopropylidene is moderately large (2.13 D with 6-31G*) reflecting donation of electrons from the pseudo-π system of the CH_2 groups to the formally vacant $2p$-orbital at the carbene center. The $2p$-electron population (4-31G) at :C is 0.051 compared with 0.047 in ethylidene (CH_3CH).

4. C_3H_2 Cyclopropenylidene (4)

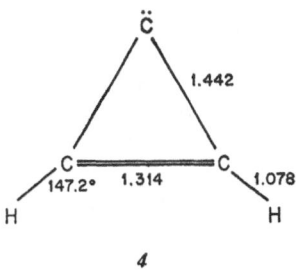

4

This molecule can be considered as a carbon-carbon double bond interacting with a carbene center. In the singlet state the carbene π-orbital is formally vacant, so there is a total of two π-electrons. This may be expected to lead to a stabilized aromatic-type of molecule [49], isoelectronic with the symmetrical cyclopropenyl cation $C_3H_3^+$. In our study we determined the geometry assuming that the molecule is planar with C_{2v} symmetry. The aromatic character of the molecule is evidenced by a considerable shortening of the C–C single bonds (1.442 Å vs. 1.537 Å in the STO-3G structure for ethylidene) and a slight lengthening of the C=C double bond (1.314 Å vs. 1.277 Å in cyclopropene).

The energy of cyclopropenylidene is quite low. If we consider the energy of the formal reaction

$$\overset{\cdot\cdot}{\underset{HC=CH}{\diagup C \diagdown}} + 3\,CH_4 \longrightarrow 2\,CH_3\text{–}CH + CH_2\text{=}CH_2$$

which compares similar bonds in separate molecules, the 6-31G* basis predicts a value of $+5.5$ kcal/mol. The positive value indicates the stabilization due to aromatic character more than offsets the large strain energy inherent in the three-membered ring. In fact, a more extensive study of C_3H_2 isomers [50] indicates that cyclopropenylidene has an energy very similar to that of propadienylidene (4-31G) and that these are the most stable structures for C_3H_2. It is likely that preferential stabilization of the cyclic structure will lead to cyclopropenylidene representing the lowest energy isomer on the 6-31G* singlet potential surface for C_3H_2.

The electron distribution also shows evidence of aromatic character. The π-electron population of the carbene carbon is 0.362 (with the 4-31G basis) indicating moderate delocalization from the carbon-carbon double bond. This electron transfer also contributes to a substantial dipole moment (3.35 D with 6-31G*) with the carbene center at the negative end.

5. C_3H_2 Cyclopropyne (5)

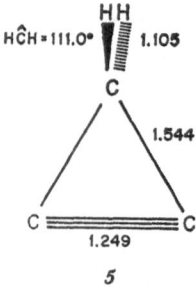

5

This is the other cyclic isomer of C_3H_2. It is expected to be highly strained because of severe angular distortion from the normal linear $-C\equiv$ geometry. The structure was determined assuming C_{2v} symmetry and leads to a triple bond slightly longer than normal. The energy is found to be 88 kcal/mol higher than that of cyclopropenylidene (with 4-31G). The bond separation energy is -150 kcal/mol indicating very high strain. In view of this considerable instability, it is possible that the ring may open spontaneously to $CH_2=C=C:$ by lowering the symmetry from C_{2v}. The transformation has not been investigated in detail but the energy of $CH_2=C=C:$ is found to be 89 kcal/mol lower than that of cyclopropyne (4-31G).[50]

Cyclopropyne has a dipole moment in the same direction as cyclopropene but the magnitude of the moment is substantially larger

(2.94 D *vs.* 0.56 D with 4-31G). The calculated π-orbital populations (4-31G) are

The dipole moment direction in cyclopropyne again is the reverse of that normally associated with a saturated hydrocarbon fragment attached to a triple bond (*e.g.* propyne).

6. C$_3$ Cyclopropynylidene (*6*)

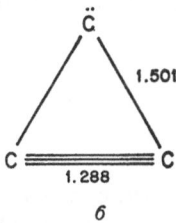

6

This species is a cyclic form of C$_3$ and may be considered as the inter-action between a triple bond and a carbene center. We assumed C_{2v} symmetry and obtained the local optimum geometry. This shows a lengthening of the triple bond and a shortening of the single bonds compared with cyclopropyne.

Although cyclopropynylidene is a local minimum in the C$_3$ surface if C_{2v} symmetry is maintained, it collapses to the linear structure, propadienediylidene (:C=C=C:) when the symmetry is lowered to C_s. Thus, cyclopropynylidene is not a stable isomer of C$_3$. The open structure is found to be lower in energy by 69.3 kcal/mol (4-31G).[51] The conversion of cyclopropynylidene to propadienediylidene proceeds *via* structures of C_s symmetry, *i.e.* the C—C single bond rather than the C≡C triple bond is broken, as would be expected.

7. C$_2$NH$_5$ Aziridine (*7*)

Aziridine, $\overline{\text{CH}_2\text{NHCH}_2}$, by analogy with open-chain saturated compounds, is expected to have a non-planar geometry for the bonds from nitrogen. This leads to a structure with a single reflection plane (C_s symmetry). In the present study, this symmetry was imposed and all remaining parameters optimized. The theoretical structure obtained is compared with the experimental one in Fig. 1 and shows generally good agreement. The angle between the N—H bond and the ring plane is calculated to be 69.7° compared to the experimental value [22] of 67.5° and other theoretical estimates [30,53] of 61° and 64°. It is interesting to note that the corresponding angle in ammonia (between N—H and the remaining NH_2 plane) is 66.4° (STO-3G) and 61.2° (experimental). Thus there is an increase in the deviation from planarity when the nitrogen is incorporated in the three-membered ring. Another feature of the experimental structure which is reproduced by the theory is a slight tilting of the methylene groups with the methylene hydrogens *cis* to N—H moving inwards.

Associated with the greater deviation from nitrogen planarity (compared to ammonia) is an increase in the barrier to inversion. We have not made a study of the C_{2v} form (with planar nitrogen) but previous theoretical calculations [30,53,54] lead to estimates in the range 15.5—18.3 kcal/mol. Experimental values of inversion barriers in simple substituted aziridines [55] are in the range 18—21 kcal/mol. For comparison, the experimental inversion barrier in ammonia [55] is 5.8 kcal/mol.

The heat of formation of aziridine is known, so it is possible to obtain an experimental bond separation energy. This is negative (− 19.3 kcal/mol after correction for zero-point vibrations) suggesting some strain but rather less than in cyclopropane. Again, this bond separation energy is fairly well described (Table 5) at the 6-31G* level.

According to both the 4-31G and 6-31G* basis sets, [56,57] aziridine is less stable than its open chain isomers, acetaldimine (CH_3—CH=NH), vinylamine (CH_2=CH—NH_2) and N-methylformaldimine (CH_2=N—CH_3).

Table 9. Energies and dipole moments for aziridine

Authors	Energy (hartrees)	Dipole moment (debyes)	Ref.
Clark	−131.8049	2.10	[30]
Bonaccorsi *et al.*	−132.6582	1.77	[35]
Lehn *et al.*	−132.9491	2.31	[53]
Basch *et al.*	−132.9726	2.40	[39]
This work (6-31G*)	−133.0352	2.09	
Expt.	−	1.89	[61]

At the 6-31G* level, acetaldimine is the most stable isomer of C_2NH_5 and it is 19.7 kcal/mol^{-1} lower in energy than aziridine.[57]

Aziridine has been examined in a number of previous *ab initio* molecular orbital studies.[30,31,34,35,38,39,53,54,59,60] Energies and dipole moments from some of these are compared with our results and with experiment [61] in Table 9. The theoretical dipole moments are all somewhat higher than the experimental value except for the result obtained by Bonaccorsi, Scrocco and Tomasi [35] with a minimal Slater basis. The latter (1.77 D) is quite close to our STO-3G value (1.82 D, Table 2).

8. C_2NH_3 1-Azirine (8) and 2-Azirine (10)

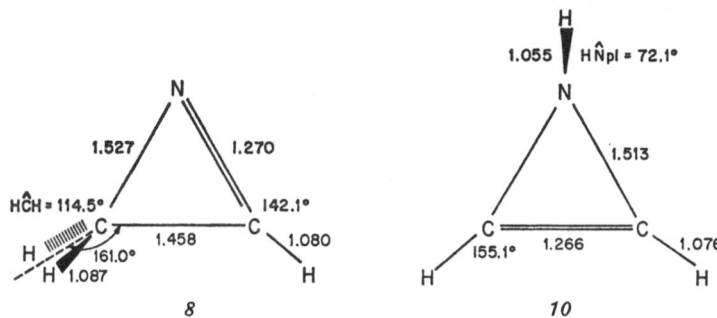

There are two possible isomeric azirines, 1-azirine (8) with a C=N bond and 2-azirine (10) with a C=C bond. They have been previously discussed in some detail by Clark.[30,38] Experimentally, only derivatives of 1-azirine are known [62] but, even for these, no structural or thermochemical data are available. For 1-azirine, we have assumed that the heavy-atom plane is a reflection plane giving overall C_s symmetry. The structure thus obtained shows a C—C bond somewhat shorter and a C—N bond somewhat longer than in acyclic molecules. For 2-azirine, we have permitted non-planarity at the nitrogen leading to C_s symmetry. The C—N bond is again found to be slightly lengthened while the C=C bond is a little shorter than in cyclopropene. In addition, the angle between the N—H bond and the ring plane is 72.1°, somewhat greater than in aziridine. Clark [30] found a similar effect which he attributed to the instability of 2-azirine in its planar form because of the antiaromatic character of its four π-electrons.

We have not made a calculation of the inversion barrier in 2-azirine. Clark [30] obtained 35 kcal/mol for this (*vs.* 15.5 kcal/mol in aziridine), again presumably because of the antiaromatic effect in the planar structure. Comparison of the two isomers shows that the energy of 1-azirine is 40.5 kcal/mol less than 2-azirine (using 6-31G*). Clark [30] obtained 27 kcal/mol for this difference. The very negative bond separation energy (Table 1) suggests that 2-azirine is unstable both because of ring strain and its unfavorable π-electron structure.

The most stable form of C_2NH_3 is acetonitrile. According to the 6-31G* basis, the 1-azirine isomer is 55.7 kcal/mol less stable than this.[57] Clark [30] has obtained 70.8 kcal/mol for this energy difference.

The dipole moments of neither 1-azirine nor 2-azirine are known experimentally. However, our values (2.56 D and 2.51 D, respectively, with 6-31G*) are quite close to the previous theoretical values obtained by Clark [30] (2.40 D and 2.50 D).

9. C_2NH_3 Aziridinylidene (9)

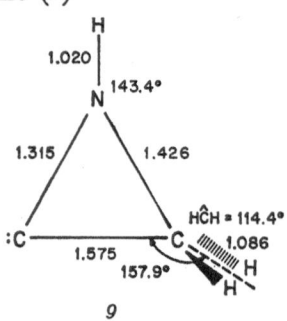

9

This third cyclic isomer of C_2NH_3 can be formally derived from aziridine (7) by abstraction of a pair of geminal hydrogens, giving a cyclic singlet carbene. However, the structure obtained shows marked differences from aziridine. Thus, the N—H bond is found to lie in the ring plane (giving the molecule C_s symmetry) and the :C—N bond is short (1.315 Å). This bond presumably acquires double bond character by virtue of conjugation between the nitrogen lone pair electrons and the formally vacant π-orbital at the carbene center. The :C—C bond is long (1.575 Å), longer than in ethylidene (1.537 Å). Energetically, aziridinylidene is predicted (by 4-31G) to lie between 1-azirine and 2-azirine.

10. C_2NH Azirinylidene (11)

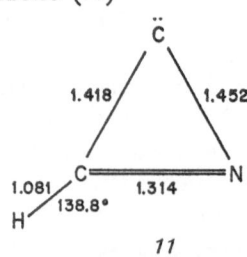

11

Abstraction of two methylene hydrogens in 1-azirine (*8*) leads to azirinylidene, which is a cyclic isomer of C_2NH with a C—H bond. A planar structure was assumed and it was found that the molecular orbital configuration involved two π-electrons giving the system some aromatic character. Donation of the π-electrons of the C=N bond into the vacant $2p\pi$ orbital at the carbene center leads to short C—C: and N—C: bonds and a slight lengthening of the C=N bond. However, this donation of π-electrons from C=N to C: is apparently not as effective as the corresponding interaction in aziridinylidene. This is reflected in the calculated N—C: lengths: 1.452 in azirinylidene compared to 3.315 in aziridinylidene (*9*).

Azirinylidene is calculated to be the most stable cyclic isomer of C_2NH. By analogy with the isoelectronic hydrocarbon cyclopropenylidene (*4*), it is possibly more stable than open chain forms. However, no theoretical studies have been made of these and experimental data appear to be lacking.

11. C_2NH Aziridinediylidene (*12*)

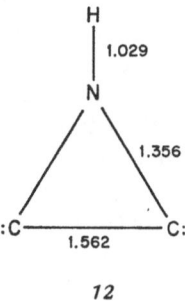

12

The other cyclic isomer of C_2NH with hydrogen attached to nitrogen may be obtained formally by replacement of the methylene group in cyclopropyne by NH. This would give a species with a C≡C bond. However, further changes in electronic structure are found leading to a system with only two π-electrons. This may be more correctly described as a dicarbene, aziridinediylidene (*12*). C_s symmetry was assumed (allowing for a non-planar nitrogen), but minimum energy was found with the N—H bond in the plane of the ring (C_{2v} symmetry). Substantial π-electron donation then takes place from N into the formally vacant $2p\pi$ orbitals on the carbon atoms. The gross π populations on these carbon atoms are 0.272 (4-31G). At the same time the C—N bonds achieve some double bond character and shorten to 1.356 Å, while the bond between the two carbene centers is quite long (1.562 Å). However, in spite of the favorable π-electron arrangement, this isomer is predicted to be 27 kcal/mol less stable than azirinylidene (*11*).

12. C_2OH_4 Oxirane (*13*)

Oxirane is the simplest cyclic ether and is well characterized experimentally. In the STO-3G geometry determination, we have assumed C_{2v}

symmetry, leading to the structure which is compared with the experimental one in Fig. 1 with moderate success. It is noteworthy that the C—C lengths in the series cyclopropane (*1*), aziridine (*7*), oxirane (*13*) decrease monotonically both theoretically (1.502, 1.491, 1.483 Å) and experimentally (1.510, 1.481, 1.471 Å). The C—O lengths in oxirane are normal.

From the known heat of formation, the bond separation energy of oxirane can be obtained. This is − 14.0 kcal/mol after zero-point correction and is partly a measure of the ring strain. By this criterion, there is a steady reduction in strain energy from cyclopropane to aziridine to oxirane. This trend is reproduced by the theory at the 6-31G* level (Table 5). However, even with d-functions included, the theoretical value for oxirane is still about 5 kcal/mol too negative.

The most stable open-chain isomer of C_2OH_4 is acetaldehyde. The energy of oxirane relative to acetaldehyde is 27.6 kcal/mol, calculated using experimental heats of formation at 0 °K (25.7 kcal/mol after zero-point correction). Theoretical values [56,57] of this energy difference are 37.8 kcal/mol (4-31G) and 30.3 kcal/mol (6-31G*). Again it is found that addition of d-functions to the basis preferentially stabilizes the cyclic isomer, leading to improved agreement for the relative energies.

Table 10. Energies and dipole moments for oxirene

Authors	Energy (hartrees)	Dipole moment (debyes)	Ref.
Clark	−151.3951	2.35	[30]
Bonaccorsi *et al.*	−152.3688	1.19	[35]
Basch *et al.*	−152.8012	2.82	[39]
This work (6-31G*)	−152.8646	2.43	
Expt.	−	1.89	[64]

There have been several previous *ab initio* molecular orbital studies of oxirane using the experimental geometry.[30,31,34,35,38,39,63] These have been mostly concerned with ionization potentials and first order properties such as multipole moments. Some details are given in Table 10. As with aziridine, all the theoretical dipole moments except those calculated with the minimal Slater basis [35] are higher than the experimental value.[64]

13. C_2OH_2 Oxiranylidene (14)

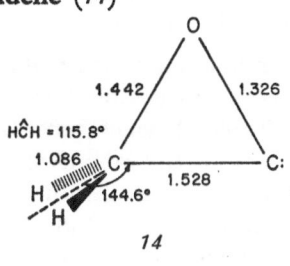

14

The most stable cyclic form of C_2OH_2 is predicted to be oxiranylidene (*14*) which is a singlet carbene formally obtained by abstraction of two geminal hydrogens from oxirane (*13*). It was assumed that the heavy atoms formed a reflection plane, giving overall C_s symmetry. The resulting structure differs from oxirane principally by a marked shortening of the O—C: bond to the carbene center (1.433 Å to 1.326 Å). This is to be expected by virtue of π-electron donation from the oxygen lone pair into the formally vacant carbene $2p\pi$ orbital giving some double bond character. At the same time there is a slight lengthening of the C—C bond. These effects are very similar to those predicted for the nitrogen analog azirinylidene (*11*).

The most stable open chain isomer of C_2OH_2 is ketene, which is predicted to be about 80 kcal/mol more stable (4-31G) [56] than oxiranylidene. Since this much lower energy can be achieved by simple cleavage of a C—O bond, it is possible that oxiranylidene may open to ketene with little activation.

14. C_2OH_2 Oxirene (15)

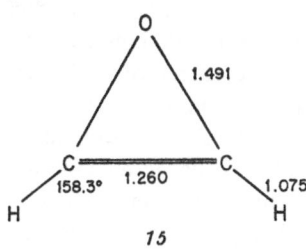

15

Oxirene (15) is the other possible cyclic isomer of C_2OH_2. It involves the interaction of an oxygen π-lone pair with a carbon-carbon double bond. However, if the molecule has C_{2v} symmetry, this is an unfavorable interaction since delocalization from O to C=C cannot take place, the vacant π^* orbital of C=C having the wrong symmetry (a_2). This situation is similar to cyclopropene where we have already noted that hyperconjugation is inhibited for the same reason. In the STO-3G structure, we have maintained C_{2v} symmetry and the lack of π-electron delocalization is shown up in the long C—O bonds (1.491 Å compared with 1.433 Å in methanol). The C—C bond is somewhat shorter than in cyclopropene.

Energetically, oxirene is found to be less stable than the oxiranylidene isomer (with 4-31G). Its low stability is also indicated by the very negative bond separation energy (Table 1) which reflects the unfavorable (antiaromatic) π-structure as well as the σ-bond strain. The total energy of oxirene is predicted at the 6-31G* level to lie 89.3 kcal/mol above that of ketene.[57] Clark,[30,38] in an earlier study of oxirene with a smaller basis also obtained a large value for this exothermicity (102.6 kcal/mol). Our calculated dipole moment (2.96 D with 6-31G*) is slightly higher than the value (2.60 D) obtained by Clark.[30]

15. C_2O Oxiranediylidene (16)

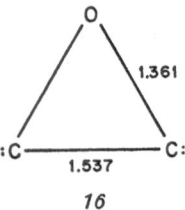

16

Replacement of the methylene group in cyclopropyne by oxygen leads to changes in electronic structure similar to those already noted for aziridinediylidene (12). The lowest energy electronic configuration has only two π-electrons and corresponds to the dicarbene oxiranediylidene (16) rather than the triply bonded species (which would have four π-electrons). There is then transfer of π-electrons from oxygen to the formally vacant $2p\pi$ orbitals on the carbene centers. In the STO-3G structure, C_{2v} symmetry was assumed and this π-electron donation (and associated partial double bond character) is reflected in short C—O bonds (1.361 Å). Oxiranediylidene is isomeric with ketenylidene (:C=C=O) and is found to be 51.4 kcal/mol less stable than the latter (4-31G).[51]

16. CN$_2$H$_4$ Diaziridine (*17*)

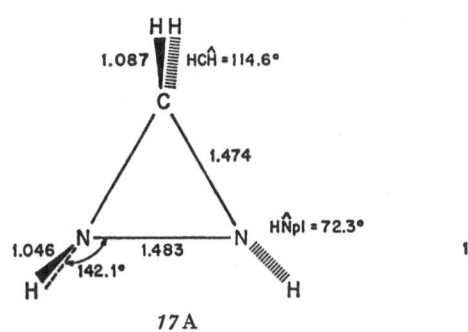
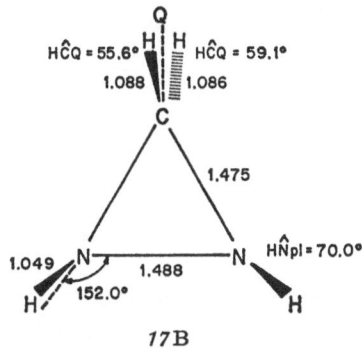

17 A   17 B

Although the parent molecule is not known, many derivatives of diaziridine have been studied experimentally and their chemistry has been reviewed.[65] Since the nitrogen valencies are non-planar, diaziridine may exist in *trans* (*17*A) and *cis* (*17*B) forms. Assumed symmetries were C_2 and C_s respectively. The STO-3G geometries show nearly equilateral triangles with N—N bonds slightly longer than in hydrazine (1.483 Å and 1.488 Å *vs.* 1.459 Å for the STO-3G length in hydrazine).

The calculations predict that *trans*-diaziridine is more stable than the *cis* isomer by 7.8 kcal/mol (4-31G). This difference is close to a previous theoretical estimate of 7.1 kcal/mol by Bonaccorsi, Scrocco and Tomasi [66] who used assumed geometries and a minimal Slater basis. The lower energy for the *trans* form can be attributed partly to the more favorable interaction between nitrogen lone pair dipoles. The theoretical bond separation energies (Table 1) at the 6-31G* level are slightly less negative than the values for cyclopropane and aziridine.

According to the 6-31G* basis, diaziridine is substantially less stable than the open chain isomer formamidine, NH$_2$—CH=NH. The calculated energy difference [57] is 43.5 kcal/mol (6-31G*).

The dipole moments of neither the *trans* nor the *cis* isomer are known experimentally. Our value for the *trans* form (1.72 D with 6-31G*) may be compared with other theoretical estimates [39,66] of 0.67 D and 1.41 D.

17. CN$_2$H$_2$ 3 H-Diazirine (*18*) and 1 H-Diazirine (*19*)

We have examined both isomeric diazirines, 3 H-diazirine with an N=N bond and 1 H-diazirine with a C=N bond. 3 H-diazirine is known and its structure has been determined by microwave spectroscopy.[24] In the STO-3G determination we have assumed C_{2v} symmetry, leading to the structure compared with experiment in Fig. 1. The N=N bondlength is

W. A. Lathan, L. Radom, P. C. Hariharan, W. J. Hehre, and J. A. Pople

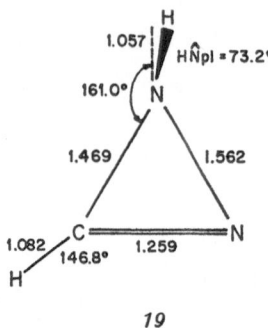

19

overestimated by about 0.04 Å, but this deficiency was also found for the N=N length in diimide (see Table 4). For 1 H-diazirine, the molecule has no symmetry. The structure is characterized by a long N—N bond and an N—H bond which makes an angle of 73.2° with the ring plane, somewhat larger than in the diaziridines (*17*). This large angle may be due to the instability of the planar form which has an antiaromatic arrangement of four π-electrons.

The theory at the 6-31G* level predicts that 3 H-diazirine (*18*) is more stable than the 1 H-isomer (*19*) by 27.9 kcal/mol. The bond separation energy (both calculated and experimental) for 3 H-diazirine is surprisingly small. Thus the 6-31G* value is − 10.7 kcal/mol, substantially smaller than the corresponding value (− 50.4 kcal/mol) for cyclopropene, which might be expected to have comparable strain energy. It seems that there is a strong stabilizing interaction between the CH_2 group and the N=N double bond in 3 H-diazirine.

The most stable isomer of CN_2H_2 is predicted to be cyanamide, $NH_2-C\equiv N$, according to the 6-31G* basis.[57] 3 H-diazirine is predicted to be 50.0 kcal/mol less stable. There have been a number of previous *ab initio* molecular orbital calculations on 3 H-diazirine.[35,36,44] These are listed in Table 11 together with calculated energies and dipole moments.

Table 11. Energies and dipole moments for 3 H-diazirine

Authors	Energy (hartrees)	Dipole moment (debyes)	Ref.
Bonaccorsi *et al.*	− 147.3772	1.29	[35]
Kochanski and Lehn	− 147.6980	2.47	[36]
Robin *et al.*	− 147.7287	2.34	[44]
This work (6-31G*)	− 147.8256	2.11	
Expt.	−	1.59	[24]

18. CN_2H_2 Diaziridinylidene (20)

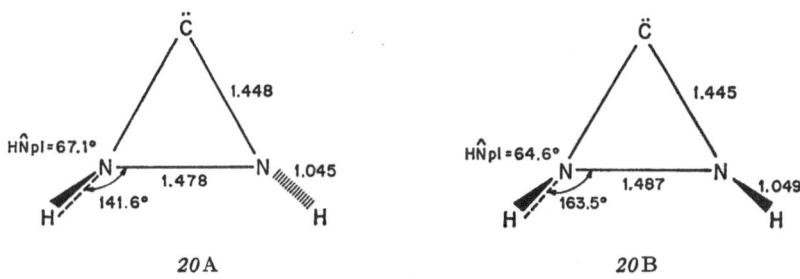

20 A   20 B

This is the other possible cyclic isomer of CN_2H_2, formed formally by removal of the two methylene hydrogens of diaziridine to give a singlet carbene structure. As with diaziridine, both *cis* and *trans* forms are possible, with C_s and C_2 symmetries respectively. The STO-3G structures both show hydrogens significantly out of the ring plane. Donation of π-electrons from the nitrogens into the vacant $2p\pi$ carbon orbital should increase the planar tendency. On the other hand, the system does have an unfavorable four π-electron arrangement in the planar form.

The 4-31G calculations show that the *trans* isomer is more stable than the *cis* by 7.7 kcal/mol. However, both are much less stable than 3 H-diazirine (by 41.9 kcal/mol for *trans*-diaziridinylidene).

19. CN_2 Diazirinylidene (21)

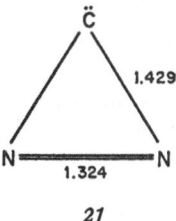

21

The cyclic form of CN_2 can be obtained formally by abstraction of two methylene hydrogens in 3 H-diazirine. This gives a singlet carbene structure symmetrically connected to an N=N bond and an aromatic system with two π-electrons. However, the energy is substantially higher (by 48.7 kcal/mol with 4-31G) than the lowest singlet of the open form NCN (cyanonitrene) which is linear with $D\infty_h$ symmetry.[51]

20. CNOH₃ Oxaziridine (22)

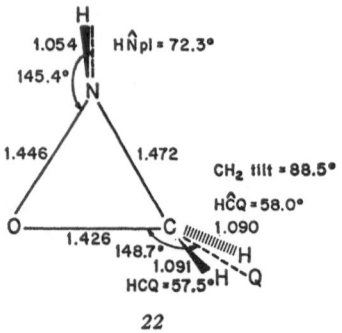

22

Only derivatives of oxaziridine are known experimentally [65] and there is therefore no experimental structure. The STO-3G geometry has no

$$H_2C—NH \quad \text{Oxaziridine}$$

element of symmetry. Notable features are a large angle (72.3°) between the N—H bond and the plane of the ring (*cf.* 69.7° in aziridine) and a twist of the CH₂ plane 1.5° away from the position perpendicular to the ring. Lehn, Munsch, Millie and Veillard [53] estimated the out-of-plane angle for N—H as 67.5° and also obtained an inversion barrier of 32.4 kcal/mol. Both these values are larger than their corresponding values for aziridine.

Oxaziridine is very much less stable than its open chain isomer, formamide. According to the 6-31G* basis [57], the energy difference is 76.9 kcal/mol. There have been several previous *ab initio* calculations on oxaziridine using assumed geometries.[53,59,60,66,67] The energies and dipole moments for these are listed in Table 12.

Table 12. Energies and dipole moments for oxaziridine

Authors	Energy (hartrees)	Dipole moment (debyes)	Ref.
Bonaccorsi *et al.*	−168.2672	2.11	[66]
Lehn *et al.*	−168.6982	3.37	[53]
Robb and Csizmadia	−168.734	3.51	[67]
This work (6-31G*)	−168.8032	3.03	

21. CNOH Oxazirine (23)

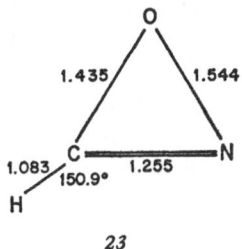

23

Oxazirine is predicted to be the most stable cyclic form of CNOH. Its π-electron structure involves an oxygen lone pair interacting with a C=N bond. This gives a four-electron antiaromatic arrangement. In the STO-3G geometry optimization procedure, the molecule was allowed to be non-planar, but the planar structure (C_s symmetry) was found to have lowest energy. The resulting geometry shows a somewhat distorted triangle with an exceptionally long N—O bond (1.544 Å). This distortion probably reduces the antiaromatic character, permitting more effective delocalization of oxygen π-electrons.

The bond separation energy is quite negative (-37.9 kcal/mol with 6-31G*), but less so than other antiaromatic molecules such as oxirene. This may partly be due to the distortion of the triangle noted above and also to a stabilizing interaction between the polar bonds. The total energy of oxazirine is predicted (by 6-31G*) to lie 97.3 kcal/mol above that of the most stable HCNO isomer, isocyanic acid.[57]

22. CNOH Oxaziridinylidene (24)

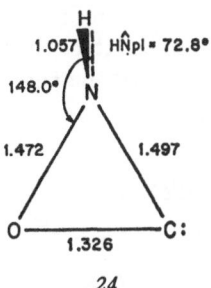

24

The other possible cyclic isomer of CNOH is obtained by removing the two methylene hydrogens of oxaziridine to give a carbene structure, oxaziridinylidene. This molecule would have four π-electrons in a planar structure. In fact, the N—H bond is found to be bent quite strongly

(72.8°) out of the ring plane, which is consistent with our results for related molecules. The C—O bond is very short (1.326 Å) indicating considerable double bond character and reflecting π-electron donation from oxygen to :C. On the other hand, the C—N bond (1.497 Å) is slightly longer than in methylamine (1.474 Å) suggesting that π-electron donation from nitrogen to :C is considerably less important. Oxaziridinylidene is found to be 20.4 kcal/mol higher in energy than oxazirine (4-31G).

23. CO_2H_2 Dioxirane (25)

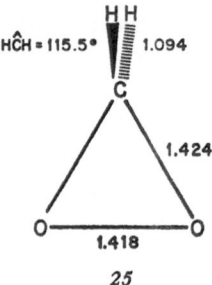

25

The cyclic peroxide, dioxirane, was assumed to have C_{2v} symmetry. The calculated O—O bond length is slightly longer than in hydrogen peroxide and may be underestimated since the STO-3G basis set underestimates this bond length in hydrogen peroxide. The bond separation energy is relatively small, − 10.7 kcal/mol with the 6-31G* flasis. This is probably due in part to the stabilizing interaction of adjacent polar bonds. For example, for the related acyclic molecule, methane diol, the stabilizing interaction between the two C—O bonds amounts to 15.5 kcal/mol (experimental).[13] Dioxirane is isomeric with formic acid and has an energy 98.4 kcal/mol higher than the latter (6-31G*).[57] Although it represents the simplest cyclic organic peroxide, it is not known experimentally at present.[68]

24. CO_2 Dioxiranylidene (26)

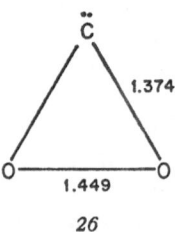

26

The geometry of this molecule was obtained assuming C_{2v} symmetry. The calculated C—O bond lengths (1.374 Å) are considerably shorter than in dioxirane (1.424 Å) though not as short as in hydroxymethylidene (1.331 Å) reflecting the extensive delocalization of the oxygen lone pairs into the vacant $2p$ orbital at :C. The carbon $2p$ electron populations in dioxiranylidene and hydroxymethylidene are 0.225 and 0.227 respectively (4-31G). Dioxiranylidene is an isomer of carbon dioxide and is calculated to be 161.3 kcal/mol less stable than the latter (4-31G).[51] If this is correct, dioxiranylidene would be unstable relative to CO + O.

25. N_3H_3 Triaziridine (27)

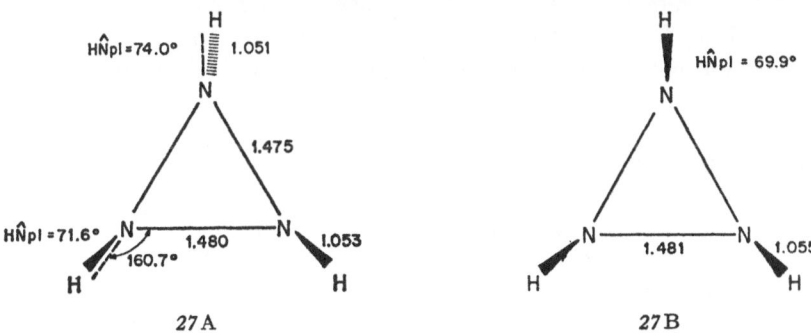

27 A   27 B

This molecule may exist as *trans* (27 A) and *cis* (27 B) isomers and the theoretical geometries of these were obtained assuming C_s and C_{3v} symmetries, respectively. The calculated geometries are quite similar to those obtained for the isoelectronic diaziridines (17).

The *trans* isomer is 17.0 kcal/mol more stable than the *cis* (4-31G), somewhat higher than the *cis-trans* energy difference in the diaziridines. The calculated bond separation energy for the *trans* isomer at the 6-31G* level is only − 8.3 kcal/mol. This suggests that there is a stabilizing interaction between the polar bonds which partially offsets the strain energy in the molecule. Such interactions have previously been noted to occur in corresponding acyclic molecules.[13] Triaziridine is predicted to be substantially less stable than its acyclic isomer, triazene NH_2—N=NH. The calculated energy difference is 40.3 kcal/mol (6-31G*).

26. N_3H Triazirine (28)

The geometry of this molecule was calculated assuming C_s symmetry. Geometric features of interest include the N=N double bond which is

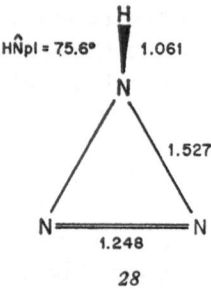

28

slightly shorter than in 3 H-diazirine (*18*), and the long N—N single bonds. The N—H bond is bent quite strongly (75.6°) out of the ring plane, a result common to many of the nitrogen containing molecules which would have four π-electrons in a planar structure. The calculated bond separation energy is slightly less negative than for 3 H-diazirine.

27. N_2OH_2 Oxadiaziridine (*29*)

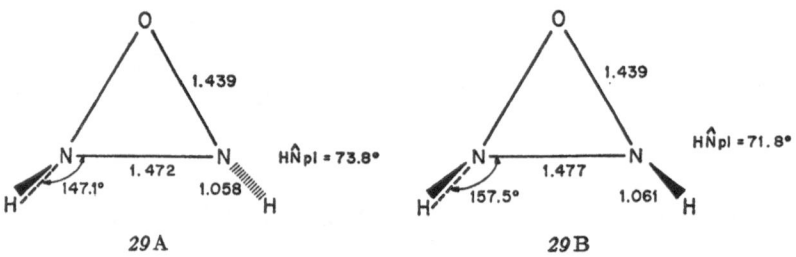

Both *trans* and *cis* isomers of oxadiziridine have been considered with C_2 and C_s symmetries, respectively. The *trans* form (*29* A) is calculated to be more stable by 8.9 kcal/mol (4-31G). There appear to be no unusual structural features. In fact, comparison of the *trans* isomers in the isoelectronic series diaziridine (*17*), triaziridine (*27*) and oxadiaziridine (*29*) shows a smooth variation in calculated geometries and in calculated bond separation energies. At the 6-31G* level, the most stable acylcic isomer of composition N_2OH_2 is hydroxydiimide, HO—N=NH. This molecule is calculated to be 57.8 kcal/mol more stable than oxadiaziridine.

28. N_2O Oxadiazirine (*30*)

This molecule is the third member of the isoelectronic series 3 H-diazirine (*18*), triazirine (*28*) and oxadiazirine (*30*). Its structure was determined

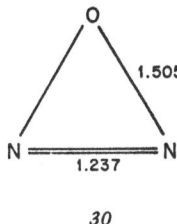

30

assuming C_{2v} symmetry. A gradual shortening of the N=N bond length in the series may be noted. Also, the calculated bond separation energies become successively less negative. Oxadiazirine itself has four π-electrons, two located on oxygen and two in the N=N double bond. Despite the four π-electron structure and the strain inherent in the three-membered ring, the bond separation energy is only -2.2 kcal/mol. There must clearly be a substantial stabilizing interaction between the adjacent polar bonds in this structure. Oxadiazirine is isomeric with nitrous oxide N=N=O and is calculated to be 75.5 kcal/mol higher in energy (4-31G).[51] Peyerimhoff and Buenker [69] have studied the variation in energy with valence angle of several configurations of nitrous oxide and also found a low energy state corresponding to this cyclic structure.

29. NO₂H Dioxaziridine (*31*)

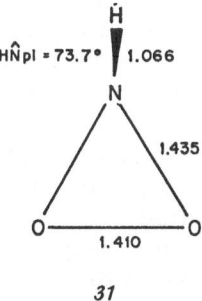

31

The theoretical structure for this molecule was obtained assuming C_s symmetry. The angle between the N—H bond and the ring plane is 73.7°, somewhat larger than in aziridine (69.7°) and oxaziridine (72.3°). Bond lengths are similar to those in related molecules. Dioxaziridine is an isomer of nitrous acid, HO—N=O and is calculated to be 84.3 kcal/mol less stable than the latter (6-31G*).[57]

30. O_3 Trioxirane (*32*)

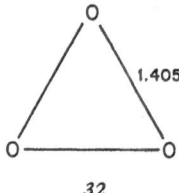

32

The structure of this molecule was initially determined assuming D_{3h} symmetry. The calculated STO-3G O—O bond length (1.405 Å) is similar to the corresponding value for hydrogen peroxide (1.396 Å). However, it is likely that both these values underestimate the actual O—O bond lengths in these molecules (see Table 4). At the STO-3G level, distortions from this D_{3h} structure obtained by lowering the symmetry to C_{2v} lead to an increase in calculated energy.

Trioxirane is a cyclic isomer of ozone but whereas trioxirane has six π-electrons, ozone has only four. The relative positions of trioxirane and ozone on the O_3 potential surface are of some interest and we have therefore carried out additional calculations in which the geometries of both species were optimized with the 4-31G basis set. This basis set gives more reasonable values of the O—O distances (1.468 Å in hydrogen peroxide [10] compared with the experimental value of 1.475 Å, 1.476 Å in trioxirane, and 1.255 Å in ozone compared with the experimental value of 1.277 Å). Using the 4-31G optimized geometries, ozone is calculated to be 18.0 kcal/mol more stable than trioxirane at the 6-31G* level. Distortions ($D_{3h} \rightarrow C_{2v}$) from the 4-31G optimized structure for trioxirane lead to an increase in (4-31G) energy, *i.e.* the cyclic structure is a local potential minimum in the 4-31G surface for O_3 under a C_{2v} symmetry constraint.

Other theoretical studies of ozone and its excited states have been previously reported and also indicate a relatively low lying state corresponding to a six π-electron cyclic structure.[70,71] In particular, recent *ab initio* calculations using the generalized valence bond and configuration interaction methods with a double zeta basis set, predict that trioxirane lies about 35 kcal/mol higher in energy than ozone but that it *is* a potential minimum.[71]

VIII. Conclusions

This survey has dealt with a large number of ring molecules of which only a few are well characterized experimentally. Many of the results are therefore of a predictive nature. However, from cases in which comparison with experiment is possible and from the general theoretical results of the study, the following conclusions emerge.

(1) Equilibrium geometries of three-membered ring molecules are reasonably well reproduced at the STO-3G level. Deviations between theoretical and experimental bond lengths and angles are comparable to those for corresponding open chain molecules.

(2) Relative energies of isomeric closed and open structures are described fairly accurately by single-determinant molecular orbital theory, provided that the basis is large enough to include *d*-functions on the heavy atoms. This suggests that changes in correlation energy may make only small contributions to the total heats of cyclization reactions.

(3) The actual energy of a three-membered ring molecule is influenced by (a) the destabilizing effect of ring-strain, (b) a π-electron effect which generally stabilizes systems with two π-electrons and destabilizes those with four, and (c) the stabilizing interaction of adjacent polar bonds.

Acknowledgements. We thank Dr. R. Ditchfield and Dr. D. P. Miller for some of the, 4-31G calculations and Dr. D. P. Miller for the complex scf program. This research was supported in part by a National Science Foundation grant, GP-25617.

IX. References

[1] Coulson, C. A., Moffitt, W. E.: Phil. Mag. *40*, 1 (1949).
[2] Walsh, A. D.: Trans. Faraday Soc. *45*, 179 (1949).
[3] These are listed in
   a) Richards, W. G., Walker, T. E. H., Hinkley, R. D.: A bibliography of *ab initio* molecular wave functions. London: Oxford University Press 1971;
   b) Radom, L., Pople, J. A.: MTP Int. rev. science, theor. chem. Ed. W. Byers Brown). London: Butterworth, in press.
[4] Roothaan, C. C. J.: Rev. Mod. Phys. *23*, 69 (1951).
[5] Hehre, W. J., Stewart, R. F., Pople, J. A.: J. Chem. Phys. *51*, 2657 (1969).
[6] Newton, M. D., Lathan, W. A., Hehre, W. J., Pople, J. A.: J. Chem. Phys. *52*, 4064 (1970).
[7] Lathan, W. A., Hehre, W. J., Pople, J. A.: J. Am. Chem. Soc. *93*, 808 (1971).
[8] Radom, L., Lathan, W. A., Hehre, W. J., Pople, J. A.: J. Am. Chem. Soc. *93*, 5339 (1971).
[9] Lathan, W. A., Hehre, W. J., Curtiss, L. A., Pople, J. A.: J. Am. Chem. Soc. *93*, 6377 (1971).
[10] Lathan, W. A., Curtiss, L. A., Hehre, W. J., Lisle, J. B., Pople, J. A.: Progr. Phys. Org. Chem., in press.
[11] Ditchfield, R., Hehre, W. J., Pople, J. A.: J. Chem. Phys. *54*, 724 (1971).
[12] a) Ditchfield, R., Hehre, W. J., Pople, J. A., Radom, L.: Chem. Phys. Letters *5*, 13 (1970);

b) Hehre, W. J., Ditchfield, R., Radom, L., Pople, J. A.: J. Am. Chem. Soc. *92*, 4796 (1970).

13) Radom, L., Hehre, W. J., Pople, J. A.: J. Am. Chem. Soc. *93*, 289 (1971).

14) Radom, L., Hehre, W. J., Pople, J. A.: J. Chem. Soc. A *1971*, 2299.

15) Hariharan, P. C., Pople, J. A.: Chem. Phys. Letters, *16*, 217 (1972).

16) Hariharan, P. C., Pople, J. A.: Theoret. Chim. Acta, *28*, 213 (1973).

17) Hehre, W. J., Ditchfield, R., Pople, J. A.: J. Chem. Phys. *56*, 2257 (1972).

18) Mulliken, R. S.: J. Chem. Phys. *23*, 1833 (1955).

19) For recent reviews, see
a) Turner, D. W., Baker, A. D., Baker, C., Brundle, C. R.: Molecular photoelectron spectroscopy. New York: Wiley 1970;
b) Brundle, C. R., Robin, M. B.: In: Determination of organic structures by physical methods, (eds. F. Nachod and G. Zucherman) Vol. III. New York: Academic Press 1971.

20) Bastiansen, O., Fritsch, F. N., Hedberg, K.: Acta Cryst. *17*, 538 (1964).

21) Kasai, P. H., Myers, R. J., Eggers, D. F., Wiberg, K.: J. Chem. Phys. *30*, 512 (1959).

22) Bak, B., Skaarup, S.: J. Mol. Struct. *10*, 385 (1971).

23) Gwinn, W. D., Le Van, W. I.: J. Chem. Phys. *19*, 676 (1951).

24) Pierce, L., Dobyns, V.: J. Am. Chem. Soc. *84*, 2651 (1962).

25) Wagman, D. D., Evans, W. H., Parker, V. B., Halow, I., Bailey, W. M., Schumm, R. H.: Selected values of physical and thermodynamic properties of hydrocarbons and related compounds, National Bureau of Standards Technical Note 270-3, Washington, D. C.: U.S. Government Printing Office 1968.

26) Benson, S. W., Cruickshank, F. R., Golden, D. M., Haugen, G. R., O'Neal, H. E., Rodgers, A. S., Shaw, R., Walsh, R.: Chem. Rev. *69*, 279 (1969).

27) Preuss, H., Diercksen, G.: Intern. J. Quantum Chem. *1*, 361 (1967).

28) Frost, A. A., Rouse, R. A.: J. Am. Chem. Soc. *90*, 1965 (1968).

29) Buenker, R. J., Peyerimhoff, S. D.: J. Phys. Chem. *73*, 1299 (1969).

30) Clark, D. T.: International Symposium on Quantum Aspects of Heterocyclic Compounds in Chemistry and Biochemistry, p. 238, Jerusalem: The Israel Academy of Sciences and Humanities 1970.

31) Snyder, L. C., Basch, H.: J. Am. Chem. Soc. *91*, 2189 (1969).

32) Newton, M. D., Switkes, E., Lipscomb, W. N.: J. Chem. Phys. *53*, 2645 (1970).

33) Klessinger, M.: Symp. Faraday Soc. *2*, 73 (1968).

34) Franchini, P. F., Zandomeneghi, M.: Theoret. Chim. Acta *21*, 90 (1971).

35) Bonaccorsi, R., Scrocco, E., Tomasi, J.: J. Chem. Phys. *52*, 5270 (1970).

36) Kochanski, E., Lehn, J. M.: Theoret. Chim. Acta *14*, 281 (1969).

37) Petke, J. D., Whitten, J. L.: J. Am. Chem. Soc. *90*, 3338 (1968).

38) Clark, D. T.: Theoret. Chim. Acta *15*, 225 (1969).

39) Basch, H., Robin, M. B., Kuebler, N. A., Baker, C., Turner, D. W.: J. Chem. Phys. *51*, 52 (1969).

40) Stevens, R. M., Switkes, E., Laws, E. A., Lipscomb, W. N.: J. Am. Chem. Soc. *93*, 2603 (1971).

41) Marsmann, H., Robert, J. B., Van Wazer, J. R.: Tetrahedron *27*, 4377 (1971).

42) Salem, L.: Pure Appl. Chem., suppl. (23rd Congress) *1*, 197 (1971).

43) Peyerimhoff, S. D., Buenker, R. J.: Theoret. Chim. Acta *14*, 305 (1969).

44) Robin, M. B., Basch, H., Kuebler, N. A., Wiberg, K. B., Ellison, G. B.: J. Chem. Phys. *51*, 45 (1969).

45) André, J. M., André, M. C., Leroy, G.: Bull. Soc. Chim. Belges *78*, 539 (1969).

46) Robin, M. B., Brundle, C. R., Kuebler, N. A., Ellison, G. B., Wiberg, K. B.: J. Chem. Phys. *57*, 1758 (1972).

[47] Benson, R. C., Flygare, W. H.: J. Chem. Phys. *51*, 3087 (1969).
[48] For a semi-empirical study of this transformation, see Dewar, M. J. S., Haselbach, E., Shanshal, M.: J. Am. Chem. Soc. *92*, 3505 (1970).
[49] For a critical discussion and leading references on the subject of aromaticity, see for example, Labarre, J.-F., Crasnier, F.: Fortschr. Chem. Forsch. *24*, 33 (1971).
[50] Hehre, W. J., Lathan, W. A., Radom, L., Pople, J. A.: Unpublished results.
[51] The energies for the acyclic structures have been taken from Ref. [52] and refer to the STO-3G optimized geometries.
[52] Lathan, W. A., Radom, L., Pople, J. A.: Unpublished results.
[53] Lehn, J. M., Munsch, B., Millie, P., Veillard, A.: Theoret. Chim. Acta *13*, 313 (1969).
[54] Veillard, A., Lehn, J. M., Munsch, B.: Theoret. Chim. Acta *9*, 275 (1968).
[55] Lehn, J. M.: Fortschr. Chem. Forsch. *15*, 311 (1970).
[56] The 4-31G energies for the acyclic molecules are taken from Ref. [13] and refer to standard geometries.
[57] The 6-31G* energies for the acyclic molecules are taken from Ref. [58] and refer to the standard geometries.
[58] Hariharan, P. C., Pople, J. A.: To be published.
[59] Levy, B., Millie, P., Lehn, J. M., Munsch, B.: Theoret. Chim. Acta *18*, 143 (1970).
[60] Vinh, J., Levy, B., Millie, P.: Mol. Phys. *21*, 345 (1971).
[61] Johnson, R. D., Meyers, R. J., Gwinn, W. D.: J. Chem. Phys. *21*, 1425 (1953).
[62] Fowler, F. W.: Advan. Heterocycl. Chem. *13*, 45 (1971).
[63] Hayes, E. F.: J. Chem. Phys. *51*, 4787 (1969).
[64] Cunningham, G. L., Boyd, A. W., Meyers, R. J., Gwinn, W. D., Le Van, W. I.: J. Chem. Phys. *19*, 676 (1951).
[65] Schmitz, E.: Advan. Heterocycl. Chem. *2*, 83 (1963).
[66] Bonaccorsi, R., Scrocco, E., Tomasi, J.: Theoret. Chim. Acta *21*, 17 (1971).
[67] Robb, M. A., Csizmadia, I. G.: J. Chem. Phys. *50*, 1819 (1969).
[68] Schultz, M., Kirschke, K.: Advan. Heterocycl. Chem. *8*, 165 (1967).
[69] Peyerimhoff, S. D., Buenker, R. J.: J. Chem. Phys. *49*, 2473 (1968).
[70] Peyerimhoff, S. D., Buenker, R. J.: J. Chem. Phys. *47*, 1953 (1967).
[71] Hay, P. J., Goddard, W. A.: Chem. Phys. Letters *14*, 46 (1972).

Received October 9, 1972

Heteroatom-Substituted Cyclopropenium Compounds

Prof. Dr. Zen-ichi Yoshida

Department of Synthetic Chemistry, Kyoto University, Yoshida, Kyoto, Japan

Contents

I. Introduction

The cyclopropenium ion system ($3C2\pi$), being the characteristic frame-work of cyclopropenium compounds, is of interest from the theoretical viewpoints set out below.

1. This electronic system is the simplest Hückel system in which 2π electrons delocalize in the three orbitals of the ring carbons.

2. Usually organic molecules are constructed by linear σ bonds, and these are the most common. However, the C—C bonds of the cyclo-propenium system are bent σ bonds (see Fig. 1) which are character-istic of small ring systems.

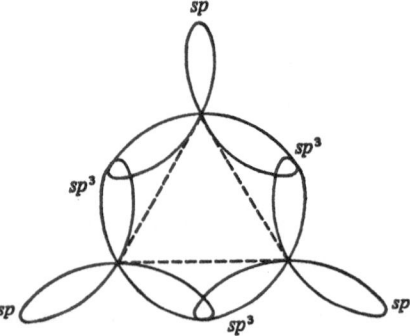

Fig. 1. Schematic drawing of the hybrid orbital of the cyclopropenium ion

3. Each ring carbon of this system has nonequivalent σ orbitals (one sp and two sp^3 orbitals[1]) as shown in Fig. 1) in contrast to the more usual organic molecules in which the hybridization of σ orbitals is equivalent. For instance, methane carbon has four equivalent sp^3 orbitals and benzene-ring carbon has three equivalent sp^2 orbitals.

4. The electronic effect of substituents in this system is not always normal. For instance, the trimethyl cyclopropenium ion (1) (pK_{R+} 6.9[2]) is more stable than the triphenylcyclopropenium ion (2) ($pK_{R+}2.8$[3]),

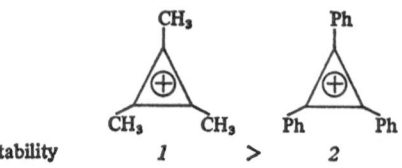

This is in marked contrast to the usual carbonium ion where the phenyl group stabilizes the carbonium ion much more than does the methyl group. (*i.e.* Ph_3C^+ is more stable than $(CH_3)_3C^+$).

Since triphenylcyclopropenyl perchlorate was first synthesized by Breslow in 1957, a number of substituted cyclopropenium compounds as well as the parent cyclopropenium compound (1967) have been reported.[4] But for the triheteroatom-substituted cyclopropenium compounds, which seem to offer a good model for investigation of the physical and chemical properties of the perturbed cyclopropenium system by heteroatoms, no compounds except the trihalocyclopropenium ion[5] were known until 1968. Since the chemistry of tri-heteroatom-substituted cyclopropenium compounds has very recently been under investigation in our laboratory, some of our work will be described here.

II. Molecular Design of Heteroatom-Substituted Cyclopropenium Compounds

To synthesize a new electronic system, tri-heteroatom-substituted cyclopropenium, we thought it useful to examine beforehand the stability of the proposed trisubstituted cyclopropenium system by quantum mechanical methods (or, in other ways). A simple Hückel molecular orbital calculation for the tri-heteroatom-substituted cyclopropenium cation (3) gave the results shown in Table 1, where q_C and p_{C-C} denote the π electron density on the ring carbon atom and the π bond order between ring carbon atoms, respectively. ΔRE (additional resonance energy) as the measure of stability of (3) is a value obtained as a resonance energy difference between (3a) and (3b).

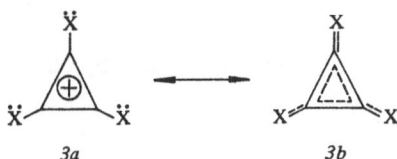

3a   *3b*

Table 1. π electron density, π bond order and additional resonance energy for the tri-substituted cyclopropenium ion, obtained by HMO

X:	H	CH_3	OH	NH_2	SH	Cl
q_C	0.667	0.734	0.832	0.862	0.806	0.764
p_{C-C}	0.667	0.633	0.584	0.569	0.592	0.618
$\Delta RE(-\beta)$	0	1.12	1.05	1.16	0.82	0.88

Table 2. Total charge (T_C, T_X), π charge (Q_C, Q_X) and π bond order (p_{C-C}) for the trisubstituted cyclopropenium ion, obtained by INDO

X:	H	CH$_3$	OH	NH$_2$
T_C	+ 0.1284	+ 0.1587	+ 0.2181	+ 0.1926
Q_C	+ 0.333	+ 0.3104	+ 0.2066	+ 0.1436
p_{C-C}	0.6667	0.6516	0.6091	0.5718
T_X	+ 0.20489	− 0.0409	+ 0.0756	− 0.3468
Q_X	−	+ 0.0054	+ 0.1152	+ 0.1897

As is seen in Table 2, INDO calculation[6] for the tri-X-substituted cyclopropenium ion (X = H, CH$_3$, NH$_2$, OH) gives similar results to HMO calculation on π charge distribution and π bond order. Both the HMO and INDO data suggest that triaminocyclo-propenium ion, if obtained, will be significantly stable among the tri-heteroatom-substituted cyclopropenium ions.

III. Synthesis of Aminocyclopropenium Compounds

To use the reaction leading to the triaminocyclopropenium ion from tetrachlorocyclopropene, which is considered to be the simplest way, seems inappropriate because the reaction of tetrachlorocyclopropene (4) with protic nucleophiles such as alcohol and water yields the ring-opened products[7] as shown in Eqs. (1) and (2).

In this case, attack of alcohol (or water) on the C—C double bond of (4) should occur non-concertedly, then further reactions might take place along Course 2 (dotted line) and Course 1 (solid line), as shown in Scheme 1. The Course 2 process should yield product (5) and the Course 1 process product (6) which might react with further alcohol to give the malonic ester.

So, in order to get the triamino (or tri-heteroatom-substituted) cyclopropenium compound, we need a $S_N 1$ or $S_N 2'$ type reaction of tetrachlorocyclopropene with amine (or nucleophile). The $S_N 1$ type

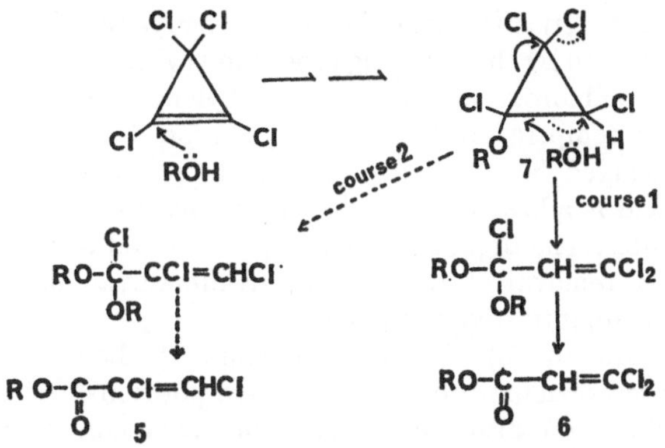

Scheme 1

reaction of tetrachlorocyclopropene leading to the triheteroatom-substituted cyclopropenium ion might occur in the case of the formation of tri-bromocyclopropenium ion by the reaction with boron tribromide[5].

Generally, amines (for example, secondary amines as protic nucleophiles) can interact with halogen compounds by donor (amines) acceptor (halogen compounds) interaction, but are not such powerful halogen-acceptors as strong Lewis acid (AlX_3, BBr_3, SbX_5). If we use excess secondary amine as both the nucleophile and halogen-acceptor, it might be expected that a concerted reaction (S_N2' type) would occur as shown in Eq. (3).

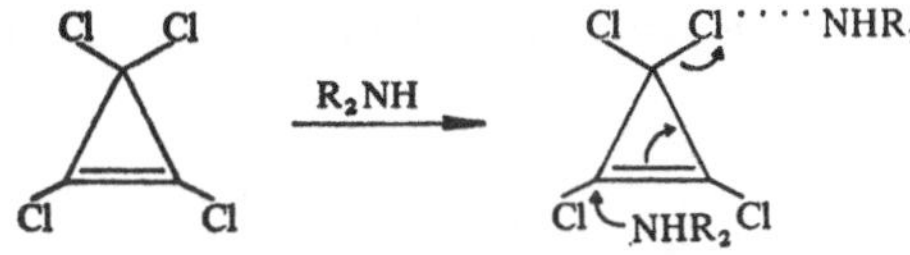

$$(3)$$

In accordance with this expectation, the reaction of tetrachlorocyclopropene with an aliphatic secondary amine (YH: dimethyl amine, piperidine, morpholine) in methylene chloride at reflux affords exclusively the triaminocyclopropenium ion in almost quantitative yield (Eq. (4)).

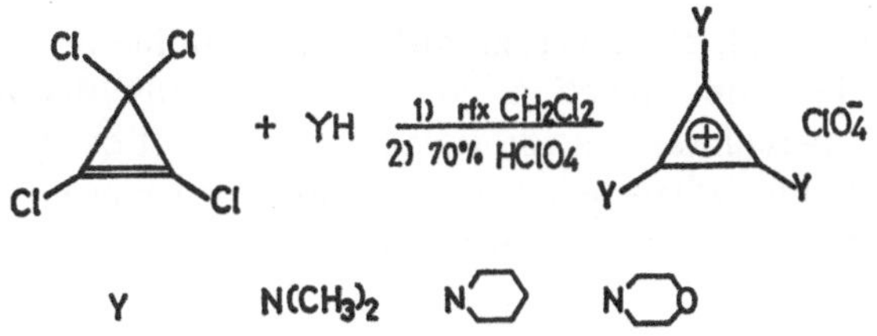

$$(4)$$

For instance, trisdimethylaminocyclopropenyl perchlorate is quantitatively obtained by the following procedures: Excess dimethylamine is added to tetrachlorocyclopropene in methylene chloride at 0 °C and stirred at this temperature for 5 h, then at room temperature for 17 h, and then refluxed for 3 h. After cooling to room temperature, 70% perchloric acid is added to the solution followed by further stirring for several minutes. The organic layer is separated and dried over sodium sulfate. After removal of the solvent, trisdimethyaminocyclopropenyl perchlorate is quantitatively obtained.

As shown in Eq. (5), N,N-dimethylaniline is allowed to react with triphenylcyclopropenium ion at the para-position[9]. The trichlorocyclopropenium ion also shows the same behavior. Even N-monoalkylaniline affords para-substituted derivatives in an electrophilic reaction such as nitration and diazo-coupling.

$$\text{(5)}$$

However, the reaction of tetrachlorocyclopropene with N-monoalkylaniline under similar conditions to those described above yields exclusively triaminocyclopropenium perchlorate. It should be noted that the reaction with diphenylamine, having two bulky phenyl groups, yields the corresponding triaminocyclopropenium perchlorate (Eq. 6)).

$$\text{(6)}$$

R   CH$_3$,   C$_2$H$_5$,   Ph

On the other hand, the reaction with diethyl amine or di-isopropyl amine under the same reaction conditions, affords diaminochlorocyclopropenium perchlorate as the isolated product (Eqs. (7) and (8)). Since diphenylamine gives exclusively triaminocyclopropenium perchlorate, the bulkiness of the alkyl group of both the amines does not seem to be the reason.

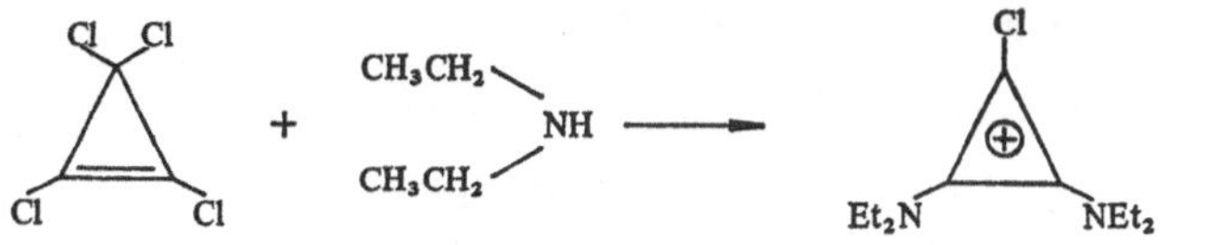

$$(7)$$

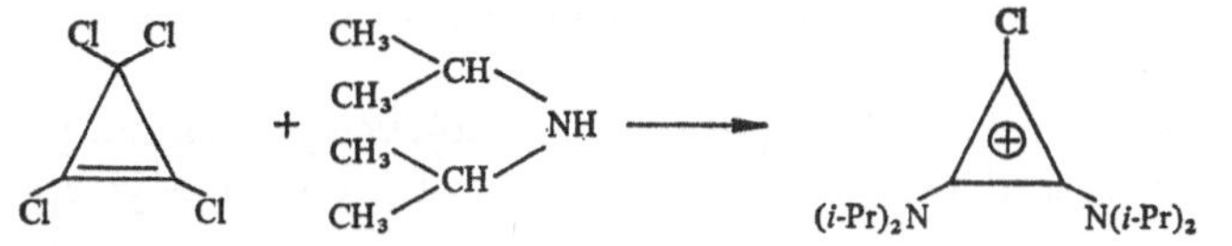

$$(8)$$

If the reaction temperature is kept at $-70\,°C$ (or lower), a secondary amine other than the two mentioned above affords the diaminochloro-cyclopropenium salt in the following yield (%): dimethylamine 10, diethylamine 56, di-*n*-propylamine 56, di-*i*-propylamine 69.

The kinetic data of the reaction have not been obtained yet because the process is so rapid. However, from the above results and discussions it might be deduced that the reaction of tetrachlorocyclopropene with a secondary amine leading to triamino- and diaminochlorocyclopropenium ions might proceed as shown in Scheme 2.

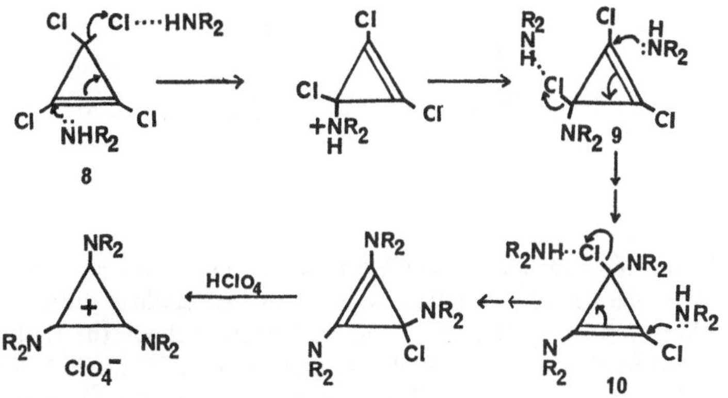

Scheme 2

An important point to emphasize is that the reactions might proceed by repeating a concerted process (S_N2' type), as is seen in (8), (9) and (10) in Scheme 2. If the reaction stops at 1-amino-2-chloro-3-amino-3-chlorocyclopropene (10) due to decreased reactivity of the cyclopropene double-bond carbon toward the secondary amine, then diaminochloro-cyclopropenium perchlorate might be obtained.

IV. Spectra and Molecular Structures

Aminocyclopropenium perchlorates are soluble in polar organic solvents, stable in the atmosphere and not hygroscopic.

The triaminocyclopropenium perchlorates in particular are very stable in water (even hot water) in contrast to the trichloro- and triphenylcyclopropenium ions, whereas the chlorides are significantly hygroscopic.

The spectral data of tris(dimethylamino)cyclopropenium perchlorate (*11*) are discussed below[8] as an example of a triaminocyclopropenium salt.

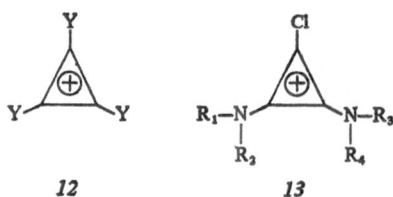

11

The nmr chemical shift (τ) in $CDCl_3$ appears at 6.84 (sharp singlet with half-width 0.3 Hz) suggesting that (*11*) cation might have the D_{3h} symmetry. Table 3 shows the nmr spectra of various triaminocyclopropenium ions (*12*) and diaminochlorocyclopropenium ions (*13*) in $CDCl_3$.

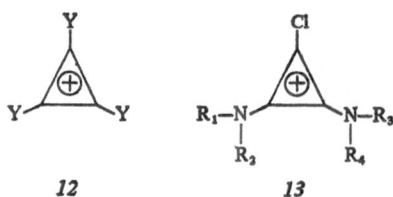

12   *13*

The nmr data for (*13*) in Table 3 indicate the existence of two kinds of alkyl groups whose magnetic environments are slightly different from each other, indicating that the rotational barrier about the C—N bond is enhanced by the increased double-bond character of the C—N linkage.

The IR spectra (KBr) of (*11*) appear at ca. 2930 (m), 1550 (vs), 1400 (s), 1280 (w), 1220 (s), 1135 (s), 1120 (m), 1090 (s), 1035 (w), 790 (m), 620 (s), 455 (w); the bands at 1090 and 620 cm^{-1} represent the absorption of ClO_4. The characteristic IR and Raman spectra of (*11*) and the band assignments based on normal coordinate analysis applying the Urey-Bradley force field are listed in Table 4. The very strong 1550 cm^{-1} bands are useful to confirm the presence of the triaminocyclopropenium ring, some of which are shown in Table 8 together with the characteristic bands for the diaminochlorocyclopropenium ring.

Table 3. Nmr spectra of triaminocyclopropenium ions (*12*) and diaminochloro-cyclopropenium ions (*13*)

Y in (*12*)	Chemical shift (τ)
$N(CH_3)_2$	6.84 (s, CH_3)
$N(CH_2)_5$	6.50 (m, 12 H, $-CH_2N$), 8.25 (m, 18 H, CH_2)
$N(CH_2)_4O$	6.30 (m, 12 H, $-CH_2O$), 5.90 (m, $12 H-CH_2N$)
$N(CH_3)Ph$	6.70 (s, 9 H, CH_3), 2.80 (m, 15 H, C_6H_5)
$N(C_2H_5)Ph$	9.00 (t, 9 H, CH_3), 6.60 (q, 6 H, CH_2), 2.70 (m, 15 H, C_6H_5)
$N(CH_2Ph)_2$	5.50 (s, 12 H, CH_2), 2.75 (m, 30 H, C_6H_5)
$N(Ph)_2$	2.60—3.60 (m, C_6H_5)

$Y(NR_1R_2 = NR_2R_3)$ in (*13*)	Chemical shift (τ) of	
	R_1 and R_3	R_2 and R_4 in (*13*)
$N(CH_3)_2$	6.66 (s, CH_3)	6.70 (s, CH_3)
$N(CH_2CH_3)_2$	6.51 (q, 4 H, CH_2)	6.54 (q, 4 H, CH_2)
	8.60 (t, 6 H, CH_3)	8.63 (t, 6 H, CH_3)
	9.00 (t, 6 H, CH_3)	9.04 (t, 6 H, CH_3)
$N(CH_2CH_2CH_3)_2$	8.20 (m, 4 H, CH_2)	8.27 (m, 4 H, CH_2)
	6.36 (t, 4 H, CH_2)	6.52 (t, 4 H, CH_2)
	8.58 (d, 24 H, CH_3)	
$N(CH(CH_3)_2)_2$	5.88 (ses, 2 H, CH)	6.14 (ses, 2 H, CH)

s: singlet d: doublet t: triplet ses: sesquitet m: multiplet

Table 4. Characteristic IR absorption of the trisdimethyl-aminocyclopropenium ion

IR (cm^{-1})	Raman (cm^{-1})	Assignments
	1985 (A_1')	C—C stretch.
1551	1551 (E')	C—C stretch.
		C—N stretch.
	1290 (A_1')	C—N stretch.
	910 (A_1')	N—Me stretch.

The vibrational spectra of the deuterated methyl derivative of (*11*), confirm that the assignments are correct. The E' ring stretching vibration appearing at 1550 cm^{-1} is the highest frequency band for cyclopropenium ions, as far as I know, although this band strongly couples with the C—N stretching vibration.

The possible molecular symmetries of the tris(dimethylamino)-cyclopropenium ion are illustrated in Fig. 2.

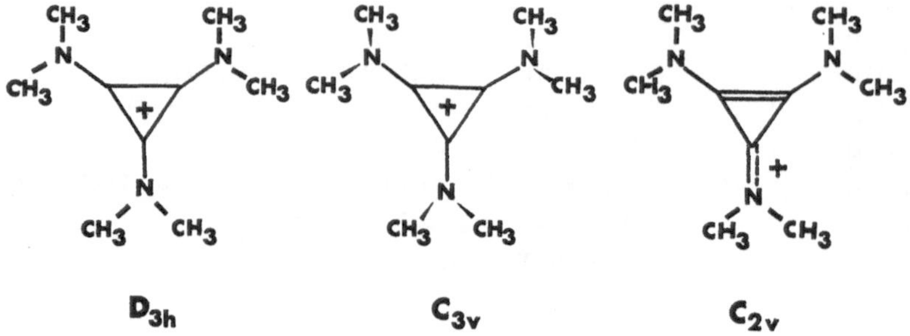

Fig. 2. Molecular symmetries for the tris(dimethylamino)cyclopropenium ion

If dimethylamino groups are coplanar with the cyclopropenium ring due to sp^2 hybridization of the amino nitrogen, the highly symmetric structure (D_{3h}) might be expected as the molecular geometry of the tris(dimethylamino)cyclopropenium ion. If the six methyl groups are out-of-plane from the C_3N_3 plane in the same direction, the C_{3v} symmetrical structure would be expected. On the other hand, if the tris (dimethylamino)cyclopropenium ion is shown by the immonium structure corresponding to triafulvene structure, the molecular symmetry of this ion is essentially approximated as C_{2v}.

Assuming one mass approximation for each methyl group, the numbers of IR- and Raman-active species for each of the molecular symmetries of the tris(dimethylamino)cyclopropenium ion are as follows:

		IR	Raman
D_{3h}	$4\,A_1' + 3\,A_2' + 7\,E' + A_1'' + 2\,A_2'' + 4\,E''$	9	15
C_{3v}	$6\,A_1 + 4\,A_2 + 10\,E$	16	16
C_{2v}	$11\,A_1 + 4\,A_2 + 5\,B_1 + 10\,B_2$	26	30

That is, the number of IR-active bands decreases with increasing molecular symmetry. Since 1) the numbers of IR-active bands in the C_{3v} and C_{2v} symmetries are much larger than those observed in spite of the one mass approximation for each methyl group, and 2), the Raman spectra of the tris(dimethylamino)cyclopropenium ion show an absorption at 1985 cm^{-1} assigned to the symmetric ring stretching vibration belonging to A' species, it is concluded that the molecular symmetry of this ion can be approximated to D_{3h}. A recent crystallographic study by Sundaralingam [10] supports this conclusion. Three skeletal vibrations of the cyclopropenium ring are classified into two normal vibrations; the A_1' (Raman active) and E' (Raman and infrared active) species. The observed frequencies of totally symmetric deformation (A_1') and degenerated deformation (E') modes are listed in Table 5.

Table 5. $\nu_{(C-C)}$ of C_3X^+ (cm^{-1})

X	A'	E'
NMe$_2$	1985	1533
CH$_3$	1880	1490
C$_6$H$_5$	1845	1411
Cl	1791	1312

It is apparent that the frequency of the A_1' mode increases with that of the E' mode. The quite stable trimethylcyclopropenium ion shows a Raman line at 1880 cm^{-1} and an infrared absorption at 1490 cm^{-1} which are assigned by normal coordinate analysis to the A_1' and E' mode vibrations of the C_3^+ ring, respectively [11]. These frequencies for $C_3Cl_3^+$ have been observed at relatively lower frequency regions (1791cm^{-1} (A$_1$)) and 1312 cm^{-1}(E')) [12]. The lowest frequency of the E' mode among cyclopropenium ions has been reported as 1276 cm^{-1} for the relatively unstable C_3H_3 ion [1]. The frequency shifts of both ring deformations due to the substituents are extremely large, as is clear from Table 5. Their shifts are possibly due to the mass effect and the electronic effect of the substituent. If the mass effect of the substituent attached to the C_3^+ core is predominant, the frequency shift to shorter wavelength will be in the following order, H < CH$_3$, N(CH$_3$)$_2$ < Cl. However, the frequency of the E' mode of C_3H_3 shows the lowest frequency among them. Therefore, it is most likely that the C$^+$ ring deformation vibrations are rather sensitive to the electronic effect of the substituent.

The normal coordinate analysis (GVFF) of the $C_3(C_6H_5)^+$ ion in the simplified model has suggested that the Raman-active line at 1845 cm^{-1} is mainly due to the C_3^+ ring deformation (80%) and is weakly coupled with the C_3^+–C < stretching vibration (20%) based on the potential energy distributions [13]. On the other hand, the IR-active band at 1411 cm^{-1} has been assigned to the C_3^+ ring deformation (41%) and the C_3^+–C < stretching vibration (51%). For discussions of the substituent effect, the Raman frequency is considered to be the preferable parameter because of the high contribution of the C_3^+ ring deformation mode.

Unfortunately, the Raman spectra of the cyclopropenium ions are less accessible than their IR spectra at the present time.

The fact that the A_1' mode of the $C_3(N(CH_3)_2)_3^+$ ion has a higher frequency than that of the $C_3(C_6H_5)_3$ ion is probably related to the difference in the C—C bond length of the C_3^+ core. According to the X-ray crystallographic study, the C—C bond length of the former ion is about 0.01 Å shorter than that of the latter ion. If this difference is significant, as has been suggested by Sundaralingam [10], the slight change in the C—C bond length of the C_3^+ core may cause a relatively large change in the vibrational spectra.

Although, the substituent effect on the cyclopropenium ring must be interpreted in terms of the force constants for a quantitative comparison, the empirical relationship between pK_R^+ and $\nu(C_3^+)E'$ is available to estimate the stability of the tri-substituted cyclopropenium ions. If the pK_R^+ values [1,2,14] are plotted against the frequencies of the E' mode, an approximate linear relationship between them is obtained, as shown in Fig. 3. For example, the pK_R^+ of trichlorocyclopropenium ion is estimated as -5, and that of triaminocyclopropenium ion [11] can be evaluated as 13 by extrapolation.

The C—C ring stretching force constants obtained by the normal coordinate analysis of tri-x-substituted cyclopropenium ions are listed in Table 6. As is seen in this Table, the force constant of the parent

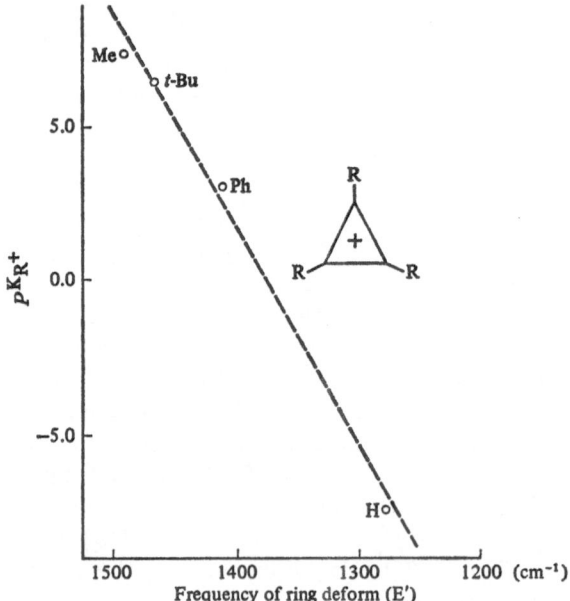

Fig. 3. Plot of pK_R^+ versus ring deformation vibration (E')

Table 6. C—C ring stretching force constants for tri-x-
substituted cyclopropenium,
ions

X	K_{C-C} (mdyne/A)
H	5.960
Cl	6.221
CH_3	6.315
$N(CH_3)_2$	6.622

cyclopropenium ion (X = H) is lower than those of the substituted ones, although it is still higher than that of benzene in which the C—C ring stretching force constant is 5.5. Since the C—C π bond orders of the cyclopropenium ring are decreased by these substituents as shown in Tables 1 and 2, the increase of the C—C ring stretching force constant might be responsible for the marked increase of C—C σ bond order.

The C—N force constant for the tris(dimethylamino)cyclopropenium ion is higher than that for the guanidinium ion [15] as shown below.

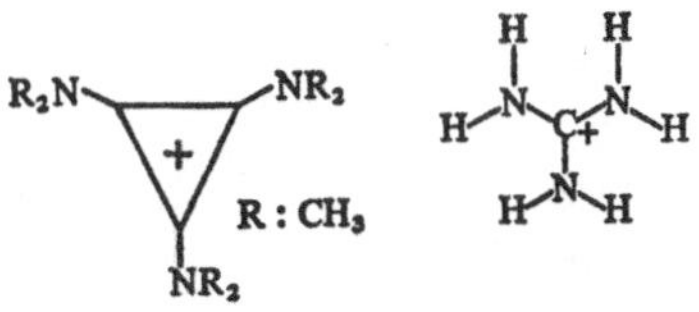

K_{C-N} 5.564 K_{C-N} 5.296

The higher C—N force constant for the former might be ascribed to the sp hybrid nature of the orbital of the ring carbon atom used for C—N bond formation as against the sp^2 hybrid nature of the corresponding orbital in the latter case.

The electronic transition (energy and oscillator strength) calculated for tri-X-substituted cyclopropenium ion by INDO [16] is set out in Table 7. The $^1\pi$-$^1\pi^*$ transition energy calculated for the triaminocyclopropenium ion by the $VI/_2$ method (modified $P \cdot P \cdot P \cdot$) is 5.917 eV(CT), which accords with the value obtained by INDO (Table 5). From these theoretical treatments of the electronic transition, the observed spectrum ($\lambda_{max} = 233$ nm ($E = 5.33$ eV), log $\varepsilon = 4.22$) for tris(dimethylamino) cyclopropenium ion is assigned to an intramolecular charge-transfer band from the amino group to the cyclopropenium ring. Table 8 shows the UV spectral data of the triaminocyclopropenium (12) and diamino-chlorocyclopropenium ions (13).

Table 7. Electronic transition energy (E) and oscillator strength (f) for tri-x-substituted cyclopropenium ions, calculated by INDO

	H	CH_3	OH	NH_2
$^1\pi^{-1}\sigma^*$				
E (eV)			6.288	5.26
f			0.0	0.0
$^1\pi^{-1}\pi^*$				
E (eV)	8.653	8.468	6.374 (CT)	5.877 (CT)
f	0.485	0.872	0.865	1.079

(CT) denotes the transition due to charge transfer from X to the cyclopropenium ring.

Table 8. Infrared and ultraviolet spectra of (12) and (13)

(12), (13)	Characteristic IR spectra (cm^{-1})[1]		UV spectra (nm[2]) λ_{max} (log ε)
Y in (12)			
$N(CH_3)_2$	1551		233 (4.22)
$N(CH_2)_5$	1533		237 (4.40)
$N(CH_2)_4O$	1525		234 (4.37)
$N(CH_3)Ph$	1509		230 (4.18), 279 (4.45)
$N(C_2H_5)Ph$	1506		230 (4.18), 277 (4.30)
$N(CH_2Ph)$	1520		210 (end), 248 (4.37)
NPh_2	1448		267 (4.46), 312 (4.24)
$NR_1R_2 = NR_2R_3$ in (13)			
$N(CH_3)_2$	1952	1642	206.5 (4.07)
$N(CH_2CH_3)_2$ IIb	1941	1607	212 (4.14)
$N(CH_2CH_2CH_3)_2$ IIc	1940	1601	213 (4.17)
$N(CH(CH_3)_2)_2$ IId	1930	1585	210.5 (4.24)

[1] Measured in KBr pellet.
[2] Measured in methanol.

From the INDO investigation for the tri-x-substituted cyclopropenium ion, the orbital energies of the lowest unoccupied molecular orbital (LUMO) and the highest occupied molecular orbital (HOMO) are given as follows:

Orbital energy (eV)

	H	CH_3	OH	NH_2
LUMO (π^*)	-6.125	-5.206	-5.224	-3.738
HOMO (π)	-21.902	-19.912	-17.277	-14.849

And the electronic effect of the substituent on the ring carbon is as indicated below:

Electronic effect of substituent on the ring carbon

	CH_3	OH	NH_2
π-Conjugative	$+0.005$	$+0.115$	$+0.190$
σ-Inductive	$+0.152$	-0.012	-0.050

It should be noted that in the CH_3 group the σ-inductive effect is larger than the π-conjugative effect, in contrast to that in the benzene system in which $\sigma_R > \sigma_I$. This novel substituent effect explains why the trialkylcyclopropenium ion is more stable than the triphenylcyclopropenium ion. On the other hand, the aminogroup has a large electron-donating π-conjugative effect as compared with its electron-drawing σ-inductive effect, in accord with its stability.

The molecular geometry of the tris(dimethylamino)cyclopropenyl cation as determined by X-ray diffraction is shown in Fig. 4. As is seen in this figure, the cyclopropenyl ring is a regular triangle with a C—C bond length of 1.363 Å. This value is significantly less than the C—C bond distance of 1.398 Å in benzene. Furthermore, it should be noted that this C—C bond length is about 0.010 Å shorter than that in the triphenylcyclopropenyl cation (1.373 Å). Since the π bond order decreases with the increasing π-conjugative effect of the substituent, the C—C bond shortening seems to imply that the C—C σ bond order is increased by the amino group. The average exocyclic C—N bond distance of 1.333 Å is also considerably shorter than the normal C—N single bond distance of 1.47 Å but close to the C—N double bond distance of 1.29 Å. This indicates that a strong delocalization of the lone-pair electrons of the nitrogen atoms into the cyclopropenyl ring occurs. The N—CH_3 distance of 1.475 Å is also shorter than the normal C—N single bond distance of 1.47 Å, suggesting that the positive charge is delocalized on methyl hydrogen due to the hyperconjugation. Since the angle around the

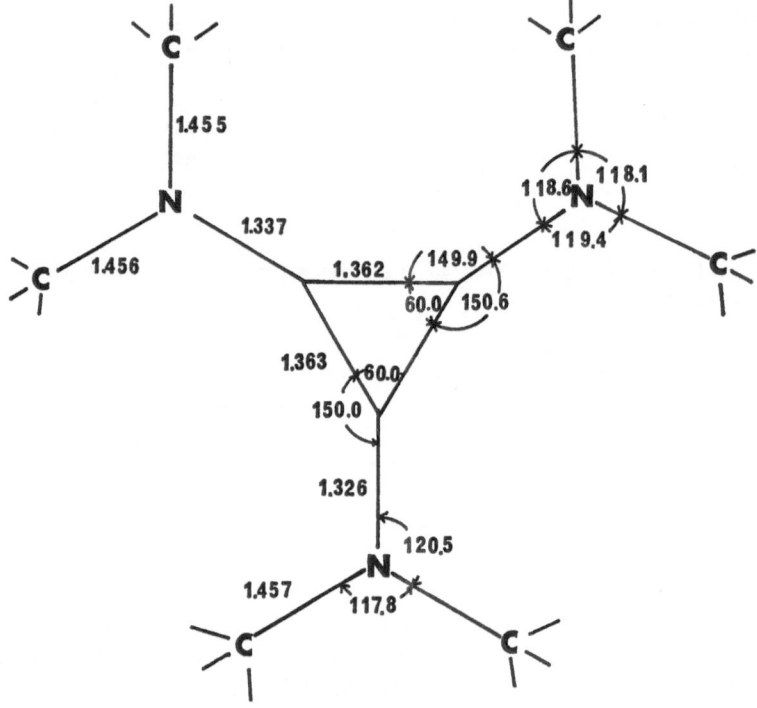

Fig. 4. Molecular structure of the tris(dimethylamino)cyclopropenium ion determined by x-ray diffraction (Sundaralingam [10])

nitrogen is almost 120°, the dimethyl amino group would appear to be coplanar with the cyclopropenyl ring.

V. Reactions of Aminocyclopropenium Ions and Diaminocyclopropenethione

A. Reactions of the Triaminocyclopropenium Ion

As shown in Eq. 9, if we can remove two electrons from the tris(dimethylamino)cyclopropenium ion (11), then we will get the hetero (3) radialene system (14) which is iso-π-electronic with the hexamethyl (3) radialene (15) synthesized by Kobrich and Heinemann [17] in 1965. In accordance with this expectation, it has been found [18] that one-electron oxidation of the diamagnetic tris(dimethylamino)cyclopropenium ion with concentrated sulphuric acid is easily effected to give the radical cation (16).

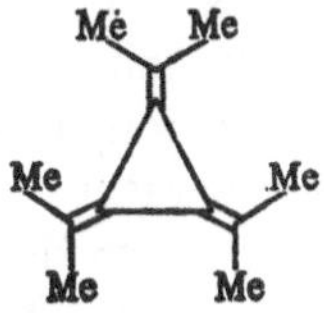

15

On the other hand, one-electron reduction of (*11*) is difficult to achieve, in contrast to the case of tri-phenylcyclopropenium ion.

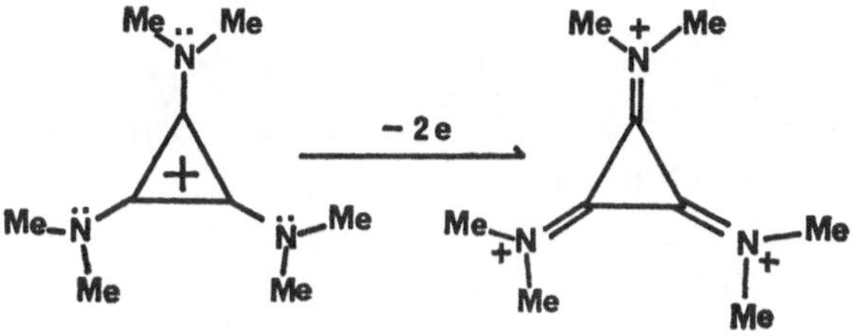

(9)

From the ESR spectrum, the spin density ϱ for the radical cation (*16*) is obtained as 0.25, which is very close to the ϱ value (0.28) of the hexamethyl (3) radialene radical anion. Although the effects of bonding and charge must be considered, the π spin population of the two radical ions is not expected to differ greatly in view of the similarity of the HMO energy diagrams of both radical ions.[18]

Although the triaminocyclopropenium ion is stable to water, even hot water, unlike the trichloro and triphenyl derivatives, this ion readily reacts with the hydroxide ion at room temperature to afford diamino-cyclopropenone (*17*) as the major product and an acrylamide derivative (*18*) as the minor product [19] (Eq. 10). On the other hand, the reaction with sodium sulfide yields diaminocyclopropenethione (*19*) and di-

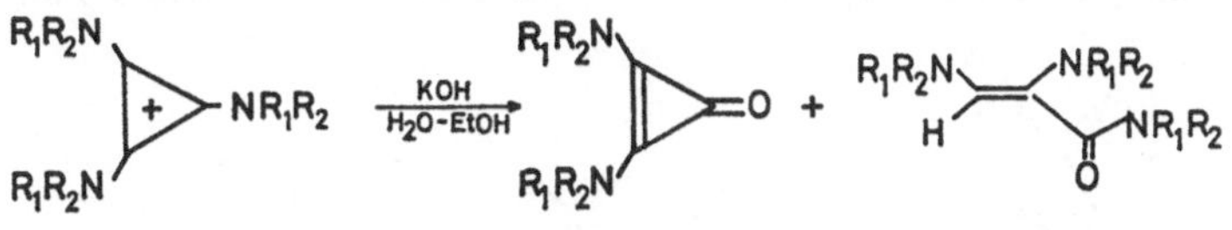

(10)

17   **18**

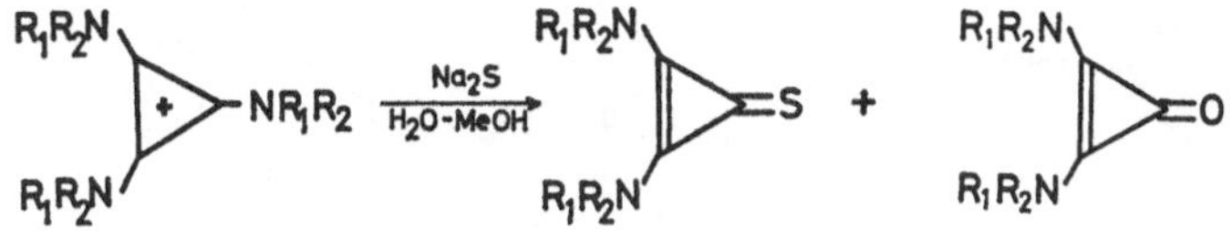

(11)

19   **17**

aminocyclopropenone (17) [20] as shown in Eq. 11. Therefore both reactions provide ways of synthesizing new compounds (17) and (19).

The triaminocyclopropenium ion is also allowed to react with potassium cyanide at room temperature to give a novel product which can be different depending on whether the substituent Y is an aromatic amino group or an aliphatic amino group.[21] If Y is an aromatic amino group, we get the product (20) in 80% yield, whereas if Y is aliphatic, for instance a piperidino group, we get the product (21) in 10% yield.

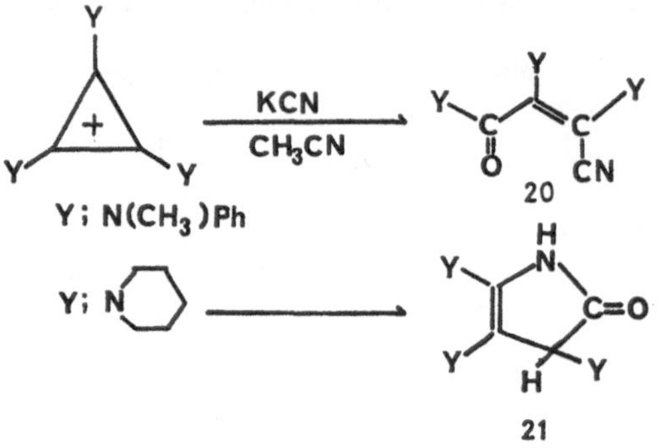

$$(12)$$

$$(13)$$

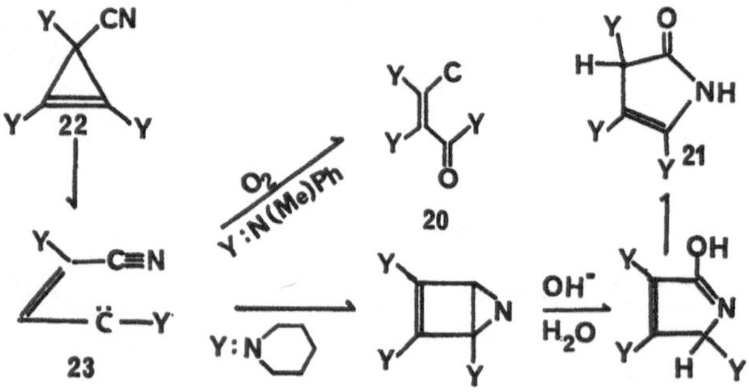

Scheme 3

The reason why different products are formed depending on the Y might be as shown in Scheme 3. In both cases, the cyclopropene derivative (22) is first formed by the nucleophilic attack of CN^-, which then might give the carbene derivative (23). This reaction is one of the features of triaminocyclopropene derivatives. When Y is an aromatic amino group, amino lone-pair electrons can conjugate with the aromatic π system, so the resultant carbene is electrophilic and can readily react with oxygen to afford the acrylamide derivative (20). On the other hand, when Y is an aliphatic amino group, the resultant carbene becomes nucleophilic, and can intramolecularly add to the CN triple bond to afford the bicyclic system which then reacts with water to give the product (*21*) (Scheme 3).

The reaction of the triaminocyclopropenium ion (*24*) with an active methylene compound in the presence of potassium *t*-butoxide affords 3,4-diaminotriafulvene (*25*) which is another new electron system.[22] The increasing solvent polarity brings about a blue shift of the longer-wavelength π-π^* absorption band (around 280—289 nm, log $\varepsilon = 4$) of (*25*). Also, the nmr chemical shift at the α-position of the piperidino group of (*25*) is nearly equal to that of the tripiperazinocyclopropenium ion (*24*). These facts suggest that the dipolar structure (*25 b*) contributes more to the ground state than the covalent structure (*25 a*).

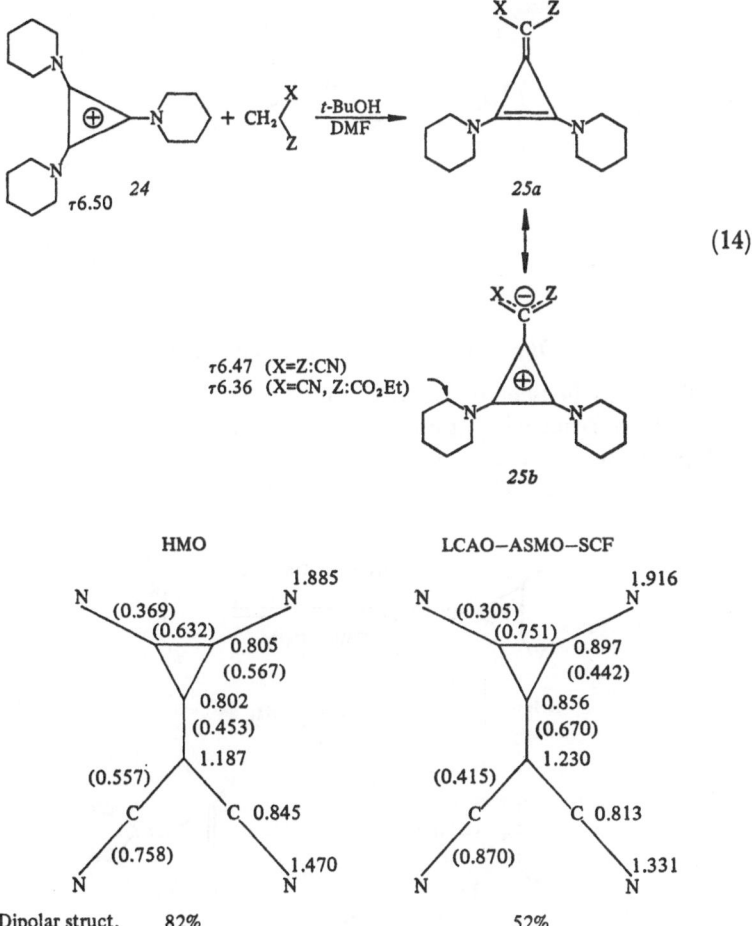

(14)

Fig. 5. Molecular diagram of 1,1-dicyano-3,4-diaminotriafulvene

The molecular diagram of 1,1-dicyano-3,4-diaminotria-fulvene obtained by HMO and LCAO-ASMO-SCF is shown in Fig. 5. The values on the atoms are the π electron densities and the values in parenthesis the π bond orders. From the molecular diagram the dipolar structure in the ground state is estimated as 82 % by HMO and as 52 % by SCFMO, which accords with the experimental results.

The reaction seems to proceed as shown in Scheme 4.

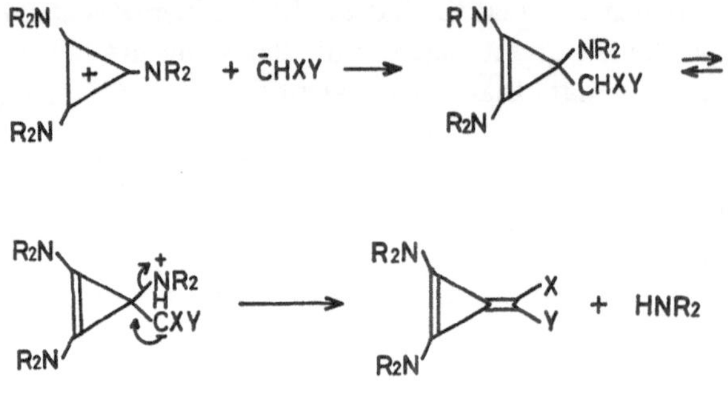

Scheme 4

B. Reactions of Diaminochlorocyclopropenium Ions

As illustrated in Scheme 5, the diaminochlorocyclopropenium ion (26) is allowed to react with ammonia to afford the antisymmetric triamino-cyclopropenium ion (27). Diaminochlorocyclopropenium perchlorate is readily hydrolyzed with potassium hydroxide in methanol-water to

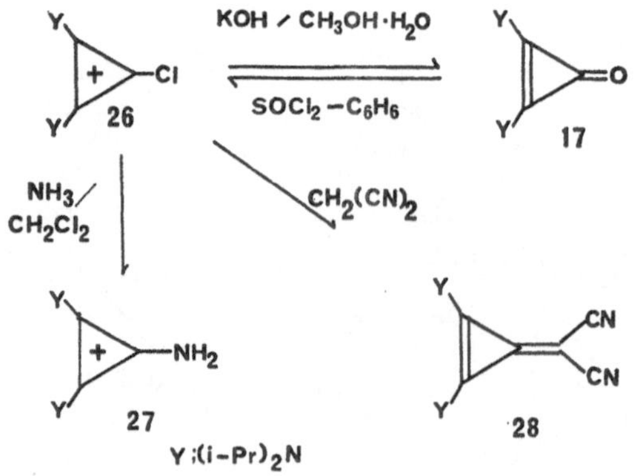

Scheme 5

afford diaminocyclopropenone (17), which is converted to the starting diaminochlorocyclopropenyl cation (26) by treatment with thionyl chloride in benzene. The diaminochlorocyclopropenium ion is also allowed to react with active methylene compounds to give 3,4-diaminotriafulvenes (28). The reactivity of the diaminochlorocyclopropenyl ion towards these reagents is higher than that of the triaminocyclopropenium ion.

Treatment of the diaminochlorocyclopropenium ion with sodium sulfide affords diaminocyclopropethione (19). On reaction with alcohol the diaminoalkoxycyclopropenium ion (29) is obtained in quantitative yield (Eq. (15)).

$$R_2N \diagup\diagdown \underset{R_2N}{\overset{+}{\triangle}}\text{-Cl} \xrightarrow[\text{rfx / CH}_2\text{Cl}_2]{\text{HOR'}} R_2N\diagup\diagdown \underset{R_2N}{\overset{+}{\triangle}}\text{-OR'}$$

29
99 %

(15)

R : n-Pr   R' : Me

The reaction with diphenylphosphine or diethylphosphine provides a synthetic pathway leading to the diaminophosphinocyclopropenium compound (30). Both the characteristic bands (A,B) indicate the existence of the cyclopropenyl ring (Eq. (16)).

$$R_2N\diagup\diagdown \underset{R_2N}{\overset{+}{\triangle}}\text{-Cl} \xrightarrow[\text{N}_2,\text{rfx/CH}_2\text{Cl}_2]{\text{HPR}_2'} R_2N\diagup\diagdown \underset{R_2N}{\overset{+}{\triangle}}\text{-PR}_2'$$

30

(16)

R	R'	Y (%)	A band	B band
			(cm⁻¹)	
ME	Ph	50	1920	1610
Et	Ph	46	1910	1580
n-Pr	Ph	86	1902	1575
i-Pr	Ph	80	1870	1550
i-Pr	Et	56	1870	1550

As reactions of the diaminochlorocyclopropenium ion other than above, the reaction with substituted aromatics will be introduced. For instance, reaction with phenol or thiophenol gives exclusively the product (*31*) or (*32*), indicating that the electrophilic attack of the diamino-chlorocyclopropenium ion takes place, not on the aromatic ring carbon, but on the hydroxyl oxygen or thiol sulfur (Eqs. (17) and (18)).

$$
\begin{array}{c}
R_2N \\
R_2N
\end{array}\!\!>\!\!\!+\!\!\!>\!\!-Cl \; + \; \overset{OH}{\underset{OCH_3}{\bigcirc}} \quad \xrightarrow[\text{Et}_3\text{N}]{\text{rfx}/\text{DCE}} \quad
\begin{array}{c}
R_2N \\
R_2N
\end{array}\!\!>\!\!\!+\!\!\!>\!\!-O\!\!-\!\!\bigcirc\!\!-OCH_3 \qquad (17)
$$

31

$$
\begin{array}{c}
R_2N \\
R_2N
\end{array}\!\!>\!\!\!+\!\!\!>\!\!-Cl \; + \; \overset{SH}{\bigcirc} \quad \longrightarrow \quad
\begin{array}{c}
R_2N \\
R_2N
\end{array}\!\!>\!\!\!+\!\!\!>\!\!-S\!\!-\!\!\bigcirc \qquad (18)
$$

R; i-Pr   32   75%

On the other hand, the reaction with anisol or dimethyl aniline afford the product (*33*) or (*34*), indicating that the reaction takes place on the phenyl carbon. The striking fact is that only the ortho-substituted compound is obtained (regio-specific reaction) (Eqs. (19) and (20)).

$$
\begin{array}{c}
R_2N \\
R_2N
\end{array}\!\!>\!\!\!+\!\!\!>\!\!-Cl \; + \; \overset{OCH_3}{\bigcirc} \quad \longrightarrow \quad
\begin{array}{c}
R_2N \\
R_2N
\end{array}\!\!>\!\!\!+\!\!\!>\!\!\underset{OCH_3}{\bigcirc} \qquad (19)
$$

33

$$
\begin{array}{c}
R_2N \\
R_2N
\end{array}\!\!>\!\!\!+\!\!\!>\!\!-Cl \; + \; \overset{N(CH_3)_2}{\bigcirc} \quad \longrightarrow \quad
\begin{array}{c}
R_2N \\
R_2N
\end{array}\!\!>\!\!\!+\!\!\!>\!\!\underset{N(CH_3)_2}{\bigcirc} \qquad (20)
$$

R; i-Pr   34

It could be a reason that the reaction intermediate (*35*) of the ortho-attack is stabilized by the electrostatic interaction, as illustrated in Scheme 6. Such an interaction is also helpful for removing chlorine from the reaction intermediate to afford (*33*) (Scheme 6).

A similar reaction assisted by electrostatic interaction is shown in Scheme 7. As is seen in this Scheme, the diaminochlorocyclopropenium ion readily reacts with sulfur ylid to give diamino triafulvenes (*35*). This reaction seems to proceed through the reaction intermediate (*36*) (Scheme 7).

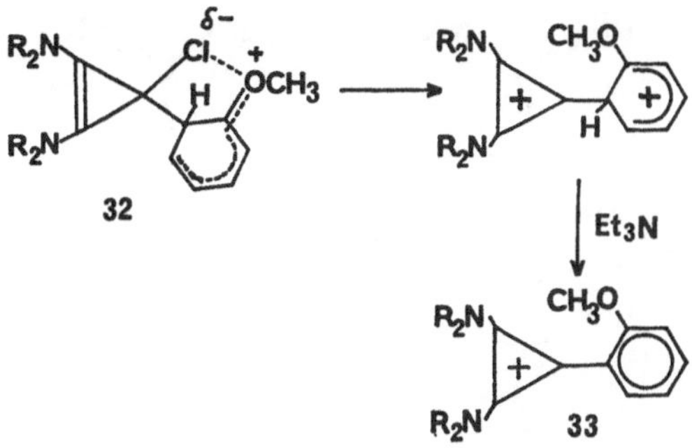

Scheme 6

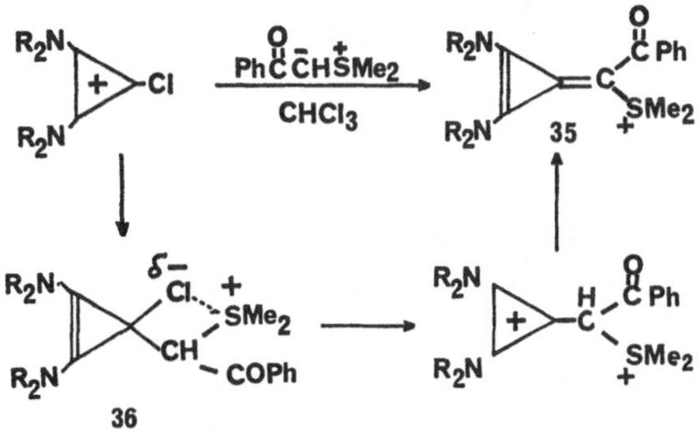

Scheme 7

C. Reactions of Diaminocyclopropenthione

Thiourea is known to behave as a C nucleophile at its sulfur atom toward an alkylating agent to yield thiuronium salt. From the similarity in the resonance structures between thiourea and diaminocyclopropenethione (*19*), as shown in Fig. 5, it might be expected that diaminocyclopropenethione when allowed to react with an alkylating agent (electrophile) will afford the diaminoalkylthiocyclopropenium ion (*37*).

In accordance with this expectation, the reaction of diamino-cyclopropenethione with dimethyl phosphite affords the diaminomethyl-thiocyclopropenium ion (*38*) (Eq. 21). This cation is also obtained by reaction with other methylating agents (*e.g.* CH_3I). The diaminothio-cyclopropenium ion (*39*), shown in Eq. 22, is obtained by the attack of 2,4-dinitrochlorobenzene on diamino-cyclopropenethione.

69

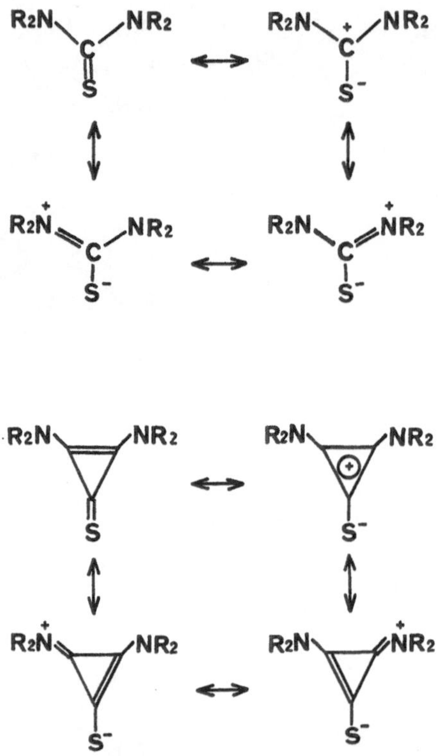

Fig. 6. Structural similarity between thiourea and diaminocyclopropenthione

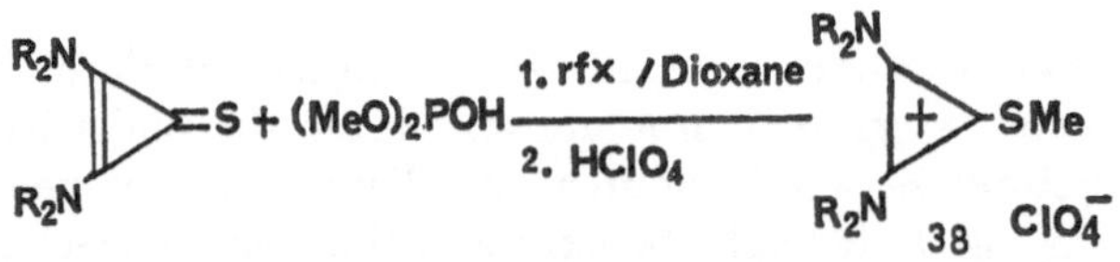

$$R; \ i-Pr \qquad 39$$

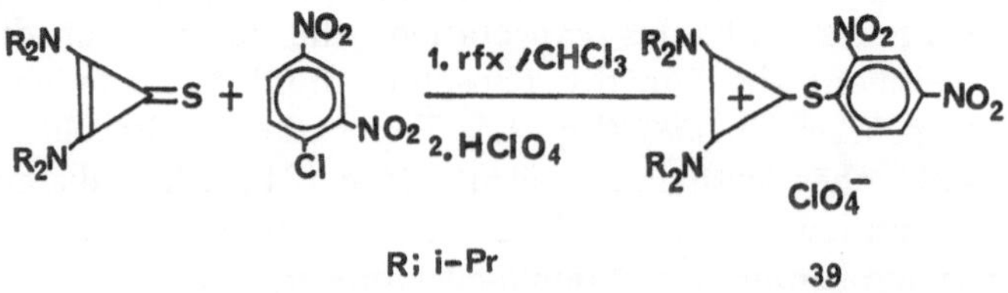

It should be noted, that p-nitrobenzenediazonium chloride is allowed to react with diaminocyclopropenthione (*19*) under Meerwein conditions to afford the heteroatom-substituted cyclopropenium compound (*40*) (Eq. 23). When the diaminochlorocyclopropenium ion is allowed to react with diaminocyclopropenthione, it gives a sulfur-bridged dication (*41*), as is shown in Eq. 24. Since diaminocyclopropenthione is synthesized from the diaminochlorocyclopropenium ion, it can be said in view of this result that the dication (*41*) is obtained from 2 moles of diamino-chlorocyclopropenium ion and 1 mol of sodium sulfide.

$$(23)$$

$$(24)$$

Interestingly, the reaction with acryl amide and benzoquinone in the presence of mineral acid affords the diaminocyclopropenium compounds (*42*) and (*43*), respectively (Eqs. 25, 26).

$$(25)$$

$$(26)$$

71

VI. Summary

In this review the chemistry of some new electron systems, "heteroatom-substituted cyclopropenium compounds", is briefly described from the viewpoints of molecular design, synthesis, molecular and electronic structures, and reactions. In each, we have encountered unique problems which will eventually produce novel fruits. In this sense the chemistry of heteroatom-substituted cyclopropenium compounds might indeed furnish topics in current chemistry and also in prospective chemistry.

Acknowledgments. Our work on the cyclopropenium compounds described above was supported by Grants from the Ministry of Education, Japan (including Grant for Scientic Research) and Mitsubishi Chemical Co., Tokyo, Japan. I would like to thank heartily many collaborators whose names appear in the References for their contributions to the work.

VII. References

1) Breslow, R., Groves, J. T.: J. Am. Chem. Soc. *92*, 984 (1970).
2) Ciabattoni, J., Nathan, III., E. C.: Tetrahedron Letters *1969*, 4997.
3) Breslow, R., Hover, H., Chang, H. V.: J. Am. Chem. Soc. *84*, 3168 (1962).
4) e.g. Closs, G. L.: In: Advances in alicyclic chemistry (Hart, H., Karabatsos, G. J., ed.), Vol. 1, p. 53. New York: Academic Press 1966.
5) West, R.: Accounts Chem. Res. *3*, 130 (1970); IUPAC, The chemistry of non-benzenoid aromatic compounds (Oki, M., ed.), p. 379. London: Butterworths 1971.
6) Yoshida, Z., Kobayashi, T.: unpublished results.
7) Tobey, S. W., West, R.: Tetrahedron Letters. *1963*, 1179.
8) Yoshida, Z., Tawara, Y.: J. Am. Chem. Soc. *93*, 2573 (1971).
9) Breslow, R.: Am. Chem. Soc. *89*, 4383 (1967).
10) Ku, A. T., Sundaralingam, M.: J. Am. Chem. Soc. *94*, 1688 (1972).
11) Yoshida, Z., Ogoshi, H., Hirota, S.: Tetrahedron Letters *1913*, 869.
12) West, R., Sado, A., Tobey, S. W.: J. Am. Chem. Soc. *88*, 2488 (1966).
13) Hofler, F., Schrader, B., Krebs, A.: Z. Naturforsch. *24a*, 1617 (1969).
14) Breslow, R., Blattiete, M.: Chem. Ind. *1958*, 1143.
15) Yamaguchi, A.: Nippon Kogaku Zasshi. *78*, 694 (1957).
16) Yoshida, Z., Kobayashi, T.: unpublished results.
17) Kobrich, G., Heinemann, H.: Angew. Chem. *77*, 590 (1965).
18) Gerson, F., Plattner, G., Yoshida, Z.: Mol. Phys. *21*, 1027 (1971).
19) Yoshida, Z., Tawara, Y., Konishi, H., Ogoshi, H.: unpublished results.
20) Yoshida, Z., Tawara, Y., Konishi, H., Ogoshi, H.: unpublished results.
21) Yoshida, Z., Tawara, Y.: unpublished results.
22) Yoshida, Z., Tawara, Y.: Tetrahedron Letters *1971*, 3603.
23) Yoshida, Z., Tawara, Y.: unpublished results.
24) Yoshida, Z., Kasai, K., Ogoshi, H.: unpublished results.
25) Yoshida, Z., Ogoshi, H., Murakami, S.: unpublished results.
26) Yoshida, Z., Konishi, H., Ogoshi, H.: unpublished results.

Received February 6, 1973

Stereochemistry and Reactivity in Cyclopropane Ring-Cleavage by Electrophiles

Prof. Charles H. DePuy

Department of Chemistry, University of Colorado, Boulder, Colorado, USA

Contents

I. Introduction

The ring-opening reactions of cyclopropane and its derivatives with electrophiles (Eq. (1)) are especially interesting from a mechanistic point of view because they involve electrophilic cleavage of a carbon-carbon single bond. They thus present problems of stereochemistry and the effect

$$\text{(structure)} + E^+ \xrightarrow{\text{SOH}} \text{(structure)} -OS + H^+ \tag{1}$$

of substituents on reaction rates for a hitherto unstudied fundamental organic reaction type. At the same time electrophilic ring-openings of cyclopropanes offer the prospect of developing into useful synthetic reactions, for if they were to proceed stereospecifically (and we shall see that they often do) they would lead to open-chain compounds containing as many as three asymmetric centers in a controlled relationship to one another. At present the reaction has only been extensively studied with two electrophiles (H+ (or D+) and Hg(OAc)₂) but its extension to other electrophiles appears imminent and, as we gain insight into the mechanistic aspects, synthetic applications will follow. It is the purpose of this review to survey the present experimental basis of our understanding of cyclopropane ring-cleavages, to point out where ambiguities and apparent contradictions lie, and to attempt to present a few unifying concepts. Since several aspects of the subject, in particular that of protonated cyclopropane itself, have been reviewed recently, this review will be selective in its coverage rather than exhaustive, emphasizing those results which bear most directly on stereochemistry and, to a somewhat lesser extent, reactivity.

II. Reaction of Substituted Cyclopropanes with Acid

As a rough generalization, cyclopropanes react with acids, HX, according to Markownikoff's Rule, the proton adding to the least substituted carbon, and the nucleophile (X⁻ or the solvent SOH) becoming attached to that carbon which can best stabilize a positive charge. As an example, 1,1,2-trimethylcyclopropane (7) opens exclusively between the C1—C3 bond when treated with fluorosulfonic acid (Eq. (2)), and 1,2-dimethylcyclopropane gives only the branched rather than the linear secondary cation under the same conditions [1]. Since we shall be most interested in events occurring at the carbon undergoing electrophilic substitution, i.e., that carbon to which the electrophile becomes attached, this means

74

$$\underset{1}{\underset{\overset{|}{H}}{\overset{CH_3}{\underset{|}{\overset{|}{\underset{C}{\overset{CH_2}{C}}}}}}} \longrightarrow \underset{2}{CH_3\overset{|}{\underset{\overset{|}{H}}{C}} - \overset{\oplus}{\underset{\overset{|}{CH_3}}{C}} - CH_3} \tag{2}$$

that we often have to work with fully substituted cyclopropanes in order to study problems of stereochemistry and reactivity at this site. Fortunately there are numerous exceptions to the strict adherence of Markownikoff's Rule, especially in less acidic solvents than FSO_3H, with reactive cyclopropanes, and in strained systems. As an example, 1,2,2-trimethylcyclopropanol (3) reacts with aqueous acid to give a mixture of $C1-C2$ and $C1-C3$ cleavage (Eq. (3)) [2].

$$\underset{3}{\underset{\overset{|}{CH_3}}{\overset{CH_3}{\underset{|}{\overset{|}{\underset{C}{\overset{CH_2}{C}}}}}}} \xrightarrow{H^+} \underset{75\%}{CH_3 - \overset{CH_3}{\underset{\overset{|}{CH_3}}{C}} - \overset{\oplus}{\underset{\overset{|}{CH_3}}{\overset{OH}{C}}}} + \underset{25\%}{CH_3 - \overset{H}{\underset{\overset{|}{CH_3}}{C}} \overset{\oplus}{\underset{\overset{|}{CH_3}}{\overset{CH_2}{C}}} - OH} \tag{3}$$

A. Stereochemistry of the Attack of the Proton

The difficulties in arriving at a uniform mechanism for the reactions of cyclopropanes with electrophiles can be recognized from the start when it is stated that there are well documented examples of cyclopropanes which react with a proton (or deuteron) exclusively with *retention* of configuration (so far the majority of cases), others which react exclusively with *inversion*, and most recently cases where a mixture of inversion and retention has been observed, with the former predominating. As a consequence, at least two competing electrophilic mechanisms must be operative, and perhaps more. The first stereochemical study was that utilizing 2-phenyl-1-methylcyclopropanol (4). This molecule was shown to undergo a mixture of $C1-C3$ and $C1-C2$ cleavage in D_2O/dioxane, the latter exclusively with *retention* [3] of configuration (Eq. (4)).

$$\underset{4}{\underset{\overset{|}{C_6H_5}}{\overset{H}{\underset{|}{\overset{|}{\underset{C}{\overset{CH_2}{C}}}}}} \overset{OH}{}} + D^+ \longrightarrow \underset{5}{\underset{\overset{|}{C_6H_5}}{\overset{H}{\underset{|}{\overset{|}{\underset{C}{\overset{CH_2}{C}}}}}} \overset{O}{\underset{\overset{|}{CH_3}}{\overset{\nearrow}{C}}}} \tag{4}$$

This result seemed reasonable enough, indicating as it did that the electrophile was attacking the electron pair of the carbon-carbon bond, and perhaps proceeding through an edge-protonated cyclopropane (*6*) as either an intermediate or transition state, with ring opening to give the very stable carbocation *7*. Since this initial observation, numerous

$$6 \qquad\qquad 7 \qquad\qquad\qquad\qquad (5)$$

other examples of the acid ring opening of cyclopropanes with retention have been recorded. Three examples, chosen to exemplify a diversity of substrate types, are given in Eqs. (6—8). In the first of these examples,

$$8 \qquad\qquad\qquad\qquad 9 \qquad\qquad\qquad (6)$$

$$10 \qquad\qquad\qquad\qquad 11 \qquad\qquad (7)$$

$$12 \qquad\qquad\qquad\qquad 13 \qquad\qquad (8)$$

8 →9, the reaction has been shown to proceed with retention by both the electrophile and nucleophile [4], the second is an anti-Markownikoff opening [5], and the third involves a bicyclobutane as the substrate [6]. Other important examples of proton ring opening with retention of configuration have been reported by Nickon [7] and by Wharton [8], among others.

Yet other cyclopropanes react with deuterons with *inversion* of configuration. The first example was provided by La Londe [9] and is

$$\text{14} \qquad \qquad \text{15} \qquad \qquad \qquad (9)$$

shown in Eq. (9). The case studied was the *exo* isomer of tricyclo [3.2.1.0^{2,4}]octane (*14*) which opens the internal cyclopropane bond with inversion by the attacking D$^+$; the deuteron becomes attached to the rear of the carbon when the C—C bond is broken. A related example has recently appeared [10]. A third example has been reported by Warnet and Wheeler [11] in which the three-membered ring in photothebainehydroquinone (*16*) opens with inversion when treated with aqueous DCl (Eq. (10)).

$$\text{16} \qquad \qquad \qquad \qquad (10)$$

Finally, Hammons and coworkers [12] showed that 1-methylnortricyclene (*17*) gives a mixture of retention and inversion when ring opened by D$^+$ in DOAc (Eq. (11)).

$$\text{17} \qquad \qquad \text{62 \% Retention} \qquad \text{38 \% Inversion} \qquad (11)$$

An even more striking example has recently been found in our laboratories, where the completely symmetrical all *cis*-1,2,3-trimethylcyclopropane (*18*) has been found to ring open with a mixture of inversion and retention [13], with the latter predominating (Eq. (12)). Thus all three possible stereochemical outcomes, complete retention, complete

77

18   Retention   Inversion

$$(12)$$

inversion and a mixture of inversion and retention, have been observed during protonation. Obviously the two pathways do not differ much in activation energy, with the retention pathway seemingly slightly favored.

B. Stereochemistry of the Reaction with the Nucleophile

Investigations into the stereochemistry of the carbon to which the nucleophile becomes attached are hampered by the fact that a carbonium ion or incipient carbonium ion is formed in the reaction, and that elimination, rearrangement and/or racemization processes frequently occur so that stereochemical information is lost. The best studies are those of La Londe [14], who has shown that various bicyclo[n.1.0]alkanes undergo internal carbon-carbon bond cleavage with the nucleophile entering predominately with *inversion* (Eq. (13)).

19   96 % Inversion   4 % Retention   (13)

One case [4] in which complete retention by the nucleophile is observed has been given in Eq. (6); and a second example of the same stereochemistry for the nucleophile has been noted by Hendrickson [15]. These results may be due to ion-pairing in the non-polar solvent used for the reaction. From the meager data available it appears that the nucleophilic stereochemistry will be similar to that expected from the corresponding solvolysis reaction under the particular conditions of the reaction.

C. Effect of Substituents on the Rate and Direction of Ring Opening

It is seldom appreciated that cyclopropanes are more reactive toward addition of acid than are olefins. Peterson [16], for example, showed that

n-butylcyclopropane reacts with trifluoroacetic acid 300 times more rapidly than do related alkenes, while LaLonde [14] observed a similar greater reactivity for his bicyclo compound *19* than for the ring-opened olefin. Cyclopropane is also known to react more readily with sulfuric acid than does ethylene [17]. McKinney [18,19] has studied the rates of ring opening of several substituted cyclopropanes in strong sulfuric acid solution at 25°. Phenylcyclopropane reacts at about 1/10 the rate of 1-methyl-1-phenylcyclopropane, which in turn reacts at nearly the same rate as cyclopropane itself. These results are in contrast to those for the hydration of alkenes, where the introduction of a phenyl group can increase the rate by a factor of 5000 [20]. Clearly the effect of substituents on the rate of ring opening of cyclopropanes is smaller than their effect on alkene addition, although additional studies are needed.

If substituents have only a small effect on the rate at which cyclopropanes are opened, their effect seems even smaller on the direction of opening in unsymmetrical cases. As implied by the Markownikoff's Rule generalization given earlier, the product from proton attack on the least substituted ring carbon generally predominates, but the predominance is usually small, except possibly in extremely strong acid. The 75:25 ratio [2] given for the ring opening of 1,2,2-trimethylcyclopropanol (Eq. (3)) is fairly typical. The same 1,1,2-trimethylcyclopropane which is reported to give 100% Markownikoff opening in FSO_3H (Eq. (2)) apparently gives a mixture of products [21] when opened in trifluoro-acetic acid (Eq. (14)), while cyclopropanols with a substituent on the 2-carbon give nearly a fifty-fifty product mixture on ring opening whether that substituent is methyl [22] (Eq. (15)) or phenyl [3] (Eq. (16)).

If there are two phenyl groups on adjacent carbons, the resultant cyclopropane bond becomes more susceptible to cleavage; 1,2-diphenyl-cyclopropanol (21, Y=H) opens to give 95% C1—C2 cleavage [23] (Eq. (17)) and this direction is negligibly effected by the presence of substituents in the C2-aryl group, or by cis, trans-isomerism. A similar

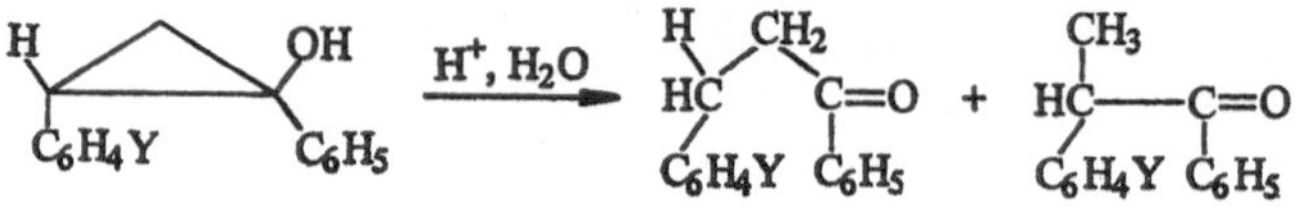

(17)

21   95 %   5 %

result is seen in the exclusive **breaking of the bond** between the phenyl groups in the compounds studied by Cristol [4] (Eq. (6)).

The relatively small effect of substituents on rate and direction of opening can be seen again in considering the anti-Markownikoff opening observed by Hendrickson [5] (Eq. (7)). Seemingly minor changes in parts of the molecule remote from the cyclopropane ring suffice to convert completely from anti-Markownikoff to Markownikoff [24] opening (Eq. (18)).

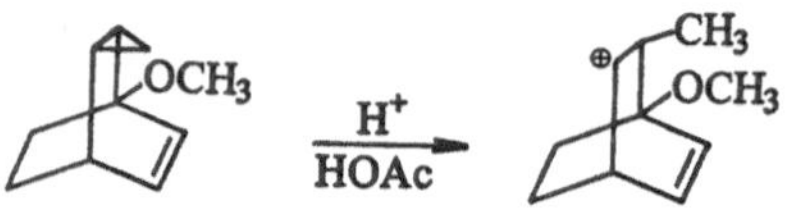

(18)

22

Cyclopropanes are, however, strongly deactivated by substitution with electron withdrawing groups [25]. One complication in attempting

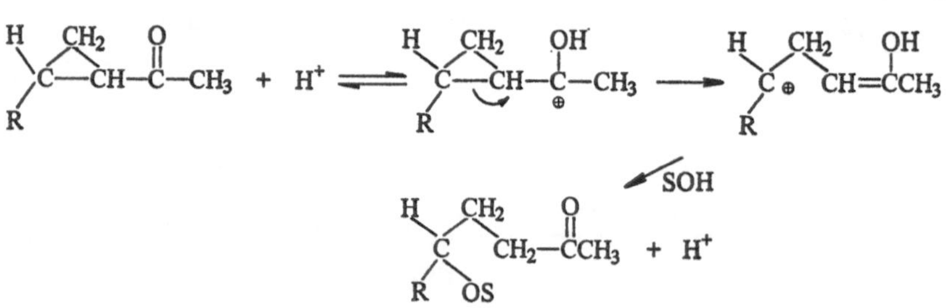

(19)

to study the ring-opening reactions of cyclopropanes substituted with carbonyl groups is that these molecules may undergo reaction by another mechanism [26] (Eq. (19)). For this reason one cannot conclude much about the direction of strictly electrophilic ring-opening in cyclopropyl ketones. Deno [27] has shown that cyclopropane carboxylic acid undergoes ring cleavage only very slowly in 98% H_2SO_4, giving rise to products resulting from both $C1-C2$ and $C2-C3$ bond cleavage (Eq. (20)).

$$
\overset{CH_2}{\underset{H_2C-CH-COOH}{\triangle}} \xrightarrow[100°, \frac{1}{2}h]{98\% \ H_2SO_4} \overset{CH_2}{\underset{O-C=O}{H_2C \quad CH_2}} + \overset{CH_2OSO_3H}{H_3C-CH-COOH} \tag{20}
$$

24%   76%

D. Protonated Cyclopropane

Protonated cyclopropanes have been extensively reviewed recently [28,29] and only the main conclusions and most important experimental data will be summarized here. The reader is referred to these reviews for greater detail. Baird and Aboderin [30] showed that when cyclopropane is treated with 8.43 M D_2SO_4 the recovered cyclopropane was partially deuterated (Eq. (21)). In later work by these same authors [17], by

$$
\overset{CH_2}{\underset{CH_2-CH_2}{\triangle}} + D^+ \longrightarrow \overset{CHD}{\underset{CH_2-CH_2}{\triangle}} + H^+ \tag{21}
$$

Deno [31] and by Lee [32,33] cyclopropane was hydrated in D_2SO_4 and the deuterium content of the n-propanol formed was examined. All three carbons contain deuterium; in 83% D_2SO_4 the deuterium distribution is statistical (28% $C-1$, 28% $C-2$, 44% $C-3$) but in 57% D_2SO_4 there is more deuterium at $C-1$ (38%) than at $C-2$ (17%). The results are summarized in Eq. (22). Adequate controls were run to ensure that the

$$
\overset{CH_2}{\underset{H_2C-CH_2}{\triangle}} \xrightarrow{D_2SO_4} \overset{CH_2-D}{\underset{H_2C-CH_2OH}{\triangle}} + \overset{CH_3}{\underset{\underset{D}{HC}-CH_2OH}{\triangle}} + \overset{CH_3}{\underset{\underset{D}{H_2C}-CHOH}{\triangle}} \tag{22}
$$

57% D_2SO_4 46%   17%   38%
83% D_2SO_4 44%   28%   28%

products are stable to the reaction conditions. These results were accommodated in terms of equilibrating edge-protonated cyclopropane intermediates (Eq. (23)). If equilibrium is established among all possible

$$\text{(23)}$$

The structures for equation (23): edge/corner protonated cyclopropanes (23, 24, 25) rearranging to deuterated propanols, plus

$$\begin{array}{c} \text{HCD—OH} \\ \text{H}_2\text{C——CH}_3 \end{array}$$

edge-protonated cyclopropanes (presumably the case in 83% D_2SO_4) a statistical mixture (3:2:2) of deuterated propanols will result, but if opening occurs before equilibration is complete, a greater than statistical amount of 1-deuteriopropanol will be formed at the expense of 2-deuterio-propanol, since formation of the latter requires two and the former only one edge-to-edge rearrangement.

The results of Baird and Aboderin cannot be accommodated by assuming that corner protonated cyclopropanes are the only reactive intermediate. If they were, 1- and 2-deuteriopropanol would of necessity be formed in equal amounts (except for a small isotope effect) no matter what the acid concentration (Eq. (24)).

$$\text{(24)}$$

Protonated cyclopropanes are also presumed intermediates in re-arrangements during the solvolysis of various compounds in which primary cations would otherwise be formed. Thus deamination of 3.3.3-trideuterio-1-aminopropane (28) leads to cyclopropane in which over half of the molecules still contain three deuterium atoms [34]; clearly this result cannot be accommodated by simple 1,3-elimination, but is understandable in terms of equilibrating protonated cyclopropanes (Eq. (25)). An alternative explanation, involving 1,3-hydrogen shifts, can be ruled

$$
\begin{array}{c}
\underset{28}{H_2C\!\!-\!\!CD_3 \atop CH_2NH_2} \longrightarrow H_2C\overset{D_2}{\underset{CH_2}{\diamond}}\!\!\cdot\!D \rightleftharpoons H_2C\overset{CD_2}{\underset{H}{\diamond}}\!\!CHD \rightleftharpoons \underline{etc.,}
\end{array}
\tag{25}
$$

$$
\overset{CD_2}{CH_2\!\!-\!\!CH_2} + D^+ \quad H_2C\overset{CD_2}{\!\!-\!\!}CHD + H^+
$$

out from the results [35] of a study of the deamination of 2,2-dideuterio-1-aminopropane (*29*). The *n*-propanol formed was examined for its deuterium content with the results shown in Eq. (26).

$$
\begin{array}{c}
\underset{29}{CH_3CD_2CH_2NH_2} \rightarrow \underset{97.9\%}{(C_2H_3D_2)CH_2OH} + \underset{1.2\%}{C_2H_5CD_2OH} \\
+ \underset{0.9\%}{(C_2H_4D)CHDOH}
\end{array}
\tag{26}
$$

If rearrangement proceeded only by 1,3-hydrogen shifts there is no way in which deuterium could appear in the 1-position. Even an unlikely series of 1,2-shifts can be ruled out by the greater amount of $C_2H_5CD_2OH$ formed than of $(C_2H_4D)CHDOH$. However, the results are easily accommodated by the intermediacy of partially equilibrating protonated cyclopropanes. A number of other solvolytic and deamination studies also support the idea that protonated cyclopropanes are reactive intermediates [28,29].

So far there is no evidence that simple alkyl substituted cyclopropanes undergo scrambling and exchange when opened with acid. On the contrary Deno [31] showed that opening of methylcyclopropane (*30*) with DCl leads exclusively to unrearranged product (Eq. (27)).

$$
\underset{30}{\overset{CH_2}{CH_2\!\!-\!\!CH\!\!-\!\!CH_3}} + DCl \longrightarrow \overset{CH_2}{\underset{D}{CH_2}\underset{Cl}{CH\!\!-\!\!CH_3}}
\tag{27}
$$

However, recent studies on long-lived carbocations in super acid solutions are beginning to give additional data. Thus Olah [36] has obtained evidence, by the use of a number of spectroscopic techniques, that the norbornyl cation is best represented as a corner protonated cyclopropane (*31*).

31

A number of studies of rearranging systems in both the gas phase [37] and in solution [1] have given evidence for the intermediacy of protonated cyclopropanes, but have not led to a distinction between the edge and corner protonated species. The gas phase proton affinity of cyclopropane has been measured, and from the data it was possible to show that protonated cyclopropane is different from n-propyl or isopropyl cation [38]. Recently, however, Saunders [39] has carried out a study of the isomerization of the isopropyl cation which indicates that corner-protonated cyclopropane is of lower energy than edge-protonated cyclopropane. The study involved an investigation of deuterium and carbon scrambling in the 1,1,1-trideuterioisopropyl cation (*32*). The ion can be formed at low temperature in SbF_5/SO_2ClF without rearrangement [40]. At 0° to $+40°$ rearrangement of both deuterium and carbon occurs [36]. As can

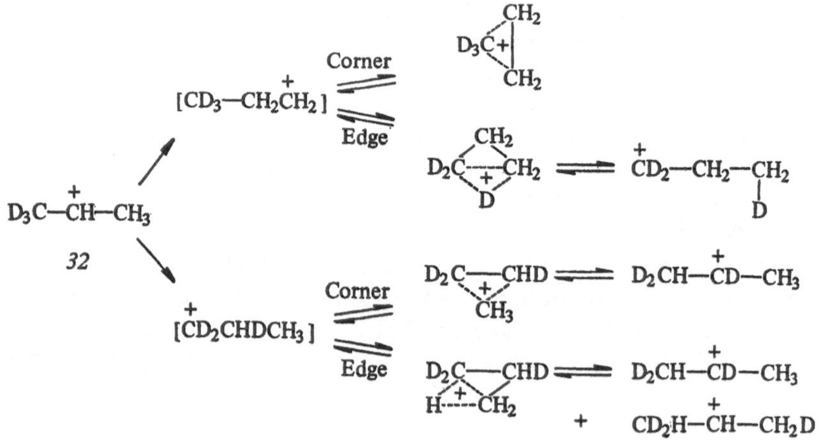

(28)

be seen from Eq. (28), if the intermediate is a corner-protonated cyclopropane, then the initially formed rearrangement product has no

1,1,3-trideuterioisopropyl cation, while a 1,3-deuterium shift is demanded by the intermediacy of the edge protonated structure. Since the initially formed rearrangement product exhibited 1,2- but not 1,3-deuterium shifts, the corner protonated structure is implicated as the intermediate in carbon atom rearrangements.

III. Cleavage of Cyclopropanes with Mercuric Salts

Lavina and coworkers [41,42] have extensively studied the reaction of cyclopropanes with mercuric acetate. In water or methanol solution the product is an organomercury alcohol or methyl ether (Eq. (29)). Studies

$$
\begin{array}{c}
\overset{\displaystyle CH_2}{\overbrace{\qquad}} \\
R—CH—CH_2
\end{array}
\; + \; Hg\,(OAc)_2 \; \xrightarrow{\;SOH\;} \;
\begin{array}{c}
\overset{\displaystyle CH_2}{\overbrace{\qquad}} \\
R—\underset{\underset{OS}{|}}{CH} \quad CH_2HgOAc
\end{array}
\qquad (29)
$$

have been made on the relative reactivity of a variety of substituted cyclopropanes toward mercuric acetate in anhydrous methanol [43]. There are surprisingly small differences in the rates of reaction; 1,1-diphenyl-, 1,2-diphenyl-(both *cis* and *trans*), ethyl- and isopropylcyclopropane all show nearly the same degree of reaction under standard conditions; *trans* 1,2-dimethylcyclopropane and phenylcyclopropane react somewhat faster. Further methyl substitution increases the rate only slightly [44]. A kinetic study of the reaction of arylcyclopropanes with mercuric acetate in acetic acid was carried out by Ouellette [45]. The reaction shows bimolecular kinetics and, surprisingly in view of the relative reactivities just quoted, gives a Hammett ϱ of -3.2 when correlated against σ^*. Under the conditions used, for example, *p*-methoxyphenylcyclopropane reacts 10,000 times more rapidly than the *m*-chloro isomer.

A. Stereochemistry of Mercuration of Cyclopropanes

The first stereochemical study of the cleavage of a cyclopropane derivative by a mercuric salt was reported by De Boer and De Puy [22]. By the series of reactions given (Eq. (30)), they showed that 1-phenyl-*cis, trans*-2,3-dimethylcyclopropanol (*33*) and 1-phenyl-*trans, trans*-2,3-dimethylcyclopropanol (*34*), as well as their methyl ethers, react with mercuric acetate in acetic acid exclusively with *inversion* of configuration at the carbon to which the mercury becomes attached. More recently

$$(30)$$

McGirk and DePuy [46,47] have reported the results of an extensive study of the stereochemistry of reaction of a variety of cyclopropane derivatives with mercuric trifluoroacetate in methanol. The stereochemistry of electrophilic attack, *i.e.*, whether the organomercurial is formed with retention or inversion, was shown to be dependent upon the substitution pattern of the cyclopropane. The results can be accommodated if it is assumed that the mercuric trifluoroacetate attacks the *least substituted* bond in the cyclopropane, followed by ring-opening in the direction of the most stable cation. If all bonds are equally substituted then attack on a *cis*-substituted bond is favored over attack on a *trans*-substituted bond. As an example, all three isomers of 1,2,3-trimethyl-1-phenylcyclopropane (*35*) are attacked by mercuric acetate exclusively with inversion of configuration, the mercuric salt attacking the 2,3-bond, which is disubstituted, rather than the trisubstituted 1,2 or 1,3 bonds. Opening in the direction of the tertiary benzylic cation results in complete *inversion* by the electrophile. It was also shown that the methyl ether is formed with complete inversion (Eq. (31)). If the 1-methyl group is

$$(31)$$

removed, giving 2,3-dimethyl-1-phenylcyclopropane, then the stereochemistry of the product is dependent upon the stereochemistry of the starting hydrocarbon. Both *trans, trans*-2,3-dimethyl-1-phenylcyclopropane (*36*) and *cis,cis*-2,3-dimethyl-1-phenylcyclopropane (*37*) give mainly *inversion* by mercury (72% and 81% respectively) since the 2,3 bond is less sterically hindered than the 1,2 or 1,3 bond. On the other hand *cis,trans*-2,3-dimethyl-1-phenylcyclopropane (*38*) gives mainly *retention* by mercury (88%) because only the 1,2 bond is *cis* substituted. In all cases the methyl ether is formed mainly with inversion. The effect

of change in the stereochemistry of the cyclopropane upon the stereochemistry of electrophilic attack is shown even more dramatically among the isomeric 2,3-dimethylcyclopropyl methyl ethers.

Since in these unsymmetrical cyclopropanes the ultimate stereochemistry seems to be determined not by a stereochemical demand of the reaction mechanism itself but rather by steric effects which determine

$$(32)$$

87

which bond is attacked, it was decided to determine the stereochemical outcome of opening *cis,cis,cis*-1,2,3-trimethylcyclopropane (*39*) with mercuric trifluoroacetate in methanol. In this molecule all bonds are equivalent so that events occurring later on along the reaction pathway must determine the stereochemistry of the reaction. The results show that for the electrophile there is a slight preference for inverson over retention (Eq. (32)), the two paths being of almost the same energy. The nucleophile is incorporated, as always, with inversion of configuration.

B. Effect of Substituents

In agreement with the stereochemical studies just reported, it has been shown that the absolute rate of attack by mercuric acetate on a cyclopropane is sensitive to the degree of substitution on the bond attacked. For example the relative rates of reaction of the three cyclopropanols [22] shown, *40*, *41* and *42*, toward mercuric acetate, are in the ratio $1:10^{-3}:10^{-6}$. In unsymmetrical cyclopropanes the direction of ring opening

is strongly influenced by steric factors [22]. This is shown in Eq. (33) in which a single methyl substituent is sufficient to induce C1—C3 cleavage

$$99\% \qquad 1\%$$

$$(33)$$

exclusively (recall that in the same system protonation gives C1—C2 cleavage to the extent of 47%). A phenyl group at the site of electrophilic attack may have some slight activating effect [22]; 2-phenyl-1-methyl-cyclopropanol (*43*) undergoes ring opening to give 25% of the product of attack at the benzylic carbon (Eq. (34)). In view of the effect of a methyl group as exemplified by Eq. (33), one would have expected less than 1% of C1—C2 cleavage in *43*. The result may indicate

$$\text{43} \xrightarrow{\text{Hg (OAc)}_2} \underset{75\%}{\text{C}_6\text{H}_5\overset{\text{CH}_2\text{HgOAc}}{\underset{\text{CH}_3}{\text{CH}-\text{C}=\text{O}}}} + \underset{25\%}{\overset{\text{CH}_2}{\underset{\text{HgOAc CH}_3}{\text{C}_6\text{H}_5\text{CH} \quad \text{C}=\text{O}}}} \quad (34)$$

that the steric effects of the phenyl group are partially offset by an electronic effect, but the data are not extensive enough to generalize.

Cyclopropanol is significantly (a factor of 10^3) more reactive toward mercuric acetate than is phenyl cyclopropane. A cyclopropyl methyl ether is less reactive than the corresponding cyclopropanol by a factor of 10—20, and the cyclopropyl acetate is decreased in reactivity by a factor of nearly 10^4. These results are fully consistent with the large ϱ value for the mercuration reaction reported by Oullette [45] (*vide supra*).

IV. Cleavage with Halogen

Reaction of cyclopropane with bromine in the presence of a Lewis acid as catalyst (FeBr$_3$, AlCl$_3$, AlBr$_3$) gives a mixture of ring-opened dibromides (1,1; 1,2; 1,3) (Eq. (35)) which are best accounted for by postulating the intermediacy of protonated cyclopropanes [48], although

$$\text{Br}^+ + \text{H}_2\text{C}-\text{CH}_2 \longrightarrow \text{H}_2\text{C}-\text{CH}_2 \rightleftharpoons \text{H}_2\text{C}-\text{CH}_2 \quad (35)$$

the data do not allow any conclusions to be drawn about the structure (edge *versus* corner) of the various intermediates involved. As is the case with mercuric acetate, reaction of bromine with substituted cyclopropanes has been found to occur with either inversion or retention at the site of electrophilic attack, depending upon the substrate structure. Thus in the cyclopropanols and derivatives studied thus far, inversion has been the sole pathway found [2], (Eq. (36)) while both retention [4] (Eq. (37)) and inversion [49] (Eq. (38)) occur in cyclopropanes investigated. In both of these latter cases the nucleophile enters with complete inversion. Bromination of bicyclo[3.1.0] hexane leads to ring-opened

$$(36)$$

(R = H, OAc)

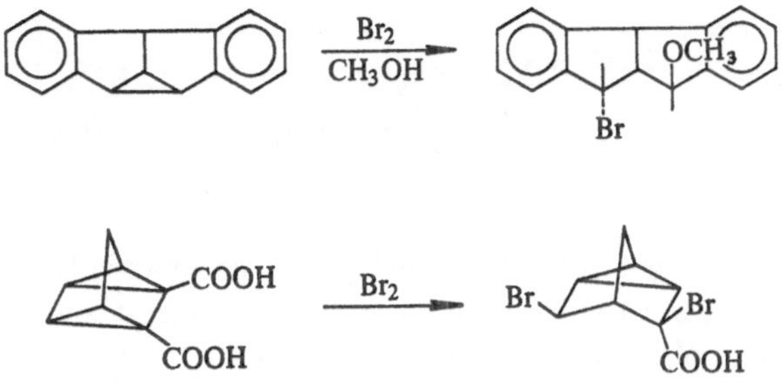

(37)

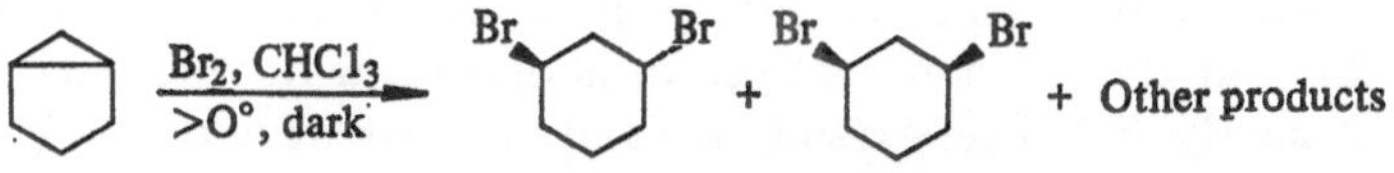

(38)

33 %   31 %

44   *45*

dibromides *44* and *45* [50]. If one assumes retention of configuration by the attacking electrophile, then the nucleophile has entered with nearly equal amounts of inversion and retention. Cyclopropane itself has been found to give only 1,3-dichloropropane upon treatment with chlorine [48], in contrast to the numerous dibromopropanes obtained on bromination. Optically active cyclopropanol *4* reacts rapidly with inversion of configuration (Eq. (39)) upon treatment with *t*-butyl hypochlorite in carbon tetrachloride [51]. On the other hand the 2,3-dimethyl-1-phenylcyclo-

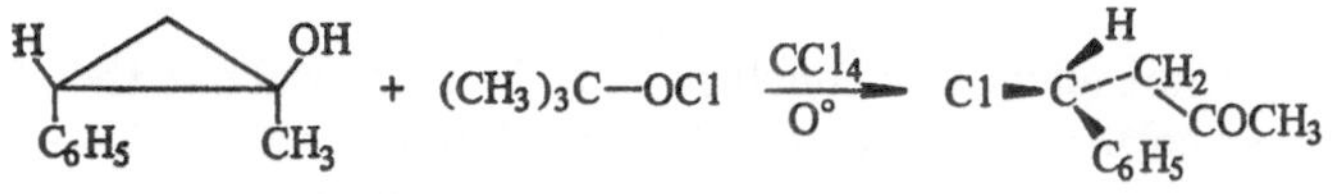

(39)

4

propanols, which react exclusively with inversion when treated with brominating agents, give equal mixtures of inversion and retention upon chlorination [2]. It was suggested [2], however, that these reactions involve attack on the O—H bond with subsequent free radical ring-opening; the great reactivity of the OH bond in cyclopropanols has been recently demonstrated [52].

V. Acylation of Cyclopropanes

Some of the most striking evidence for the intermediacy of protonated cyclopropanes comes from studies of the reaction of cyclopropane and alkylated cyclopropanes with acetyl chloride and aluminum chloride.

The product with cyclopropane itself is a mixture of 1,3(46), 1,2(47) and 1,1(48) addition [53,54], with the rearranged products predominating (Eq. (40)).

$$CH_3\overset{Cl}{\underset{}{C}}{=}O \ + \ \underset{CH_2{-}CH_2}{\overset{CH_2}{\triangle}} \ \xrightarrow{AlCl_3} \ CH_3\overset{O}{\overset{\|}{C}}CH_2CH_2CH_2Cl \ + \ CH_3\overset{O}{\overset{\|}{C}}\underset{CH_3}{\overset{}{C}}HCH_2Cl$$

46

47   (40)

$$+ \ \ CH_3\overset{O}{\overset{\|}{C}}{-}\underset{Cl}{\overset{}{C}}H{-}CH_2CH_3$$

48

In contrast to the results with DBr addition (vide supra), rearrangement (Eq. (41)) was also observed when substituted cyclopropanes were used as substrates [55].

$$CH_3\overset{+}{C}{=}O \ + \ \underset{CH_3}{\overset{CH_3}{\triangle}} \ \longrightarrow \ CH_3\overset{O}{\overset{\|}{C}}\underset{CH_3}{\overset{}{C}}H{-}\underset{CH_3}{\overset{Cl}{\overset{|}{C}}}{-}CH_3 \qquad (41)$$

VI. Reaction with Other Electrophiles

Ring-opening reactions of cyclopropanes have been observed with a variety of metallic ions. Oullette has studied the kinetics of the reaction of thallium acetate with phenylcyclopropane [56,57] (Eq. (42)).

$$YC_6H_4{-}\underset{}{\overset{CH_2}{CH{-}CH_2}} \ + \ Tl(OAc)_3 \ \longrightarrow \ [YC_6H_4\overset{OAc}{\overset{|}{C}}HCH_2CH_2Tl(OAc)_2]$$

(42)

$$YC_6H_4CH{=}CHCH_2OAc \ +$$

$$YC_6H_4\underset{OAc}{\overset{}{C}}H{-}CH_2CH_2OAc$$

The organothallium product is unstable to the reaction conditions, and decomposes to a mixture of acetate and olefin. A Hammett ϱ value was determined for the reaction (−4.3). Lead salts also react with cyclopropanes [58,59]. Recently extensive studies of the isomerization of strained cyclopropanes catalyzed by silver ions or by various transition metal catalysts have been carried out. These reactions have been the subject of an excellent recent review [60]. In addition cyclopropanes have been shown to react with palladium chloride [61], ceric ammonium nitrate [62], diborane [63,64], lithium aluminium hydride [65] and other electrophiles.

Cyclopropanes have also been shown to undergo ring-opening upon photolysis [66]. Addition of CH_3OD to dibenzotricyclo-[3.3.0.0^{2,8}] octadiene (8) gives, as the major product, retention with both electrophile and nucleophile. Cacace and coworkers [67] have carried out interesting studies of gas phase exchange reactions between cyclopropanes and the extremely strong acid helium tritide ion, formed from the β decay of molecular tritium. Hydrogen-tritium exchange is observed in cyclopropane under these conditions, and both cis- and trans-1,2-dimethylcyclopropane undergo this exchange without interconversion of the two isomers. The stereochemistry of free-radical ring-openings on substituted cyclopropanes have been shown to occur with inversion at the carbon to which the attacking radical (Cl.) becomes attached [68].

VII. Mechanisms for the Reaction of Cyclopropanes with Electrophiles

At least three general pathways need to be considered for the ring-opening reactions of cyclopropanes with electrophiles, each of which could proceed with either inversion or retention. The most straight-forward mechanism would involve a direct, single-step reaction leading to a carbonium ion, a reaction which could truly be considered an S_E2 process.

Mechanism I

Such a mechanism obviously cannot account for all of the rearrangements for the isotopic scrambling observed when cyclopropanes are reacted with electrophiles and so cannot be the sole pathway operative.

A second possible reaction path would involve the intermediacy of an edge-substituted cyclopropane, as in Mechanism II, while a corner substituted cyclopropane as in Mechanism III, is also a possibility [a].

If we postulate that all of these pathways are allowed, that each can lead either to inversion or retention, and further that edge and corner substituted cyclopropanes are in equilibrium with one another, then obviously nearly any experimental result can be accommodated. Unfortunately the situation is nearly this bad at the present time. In the following we will examine the present status of the experimental evidence.

A. Protonation

The firmest piece of experimental evidence we have is that the norbornyl cation is corner protonated and not edge protonated [36]. We therefore assume that Mechanism III can operate for protonation. But the evidence is also firm that all ring openings by a proton cannot occur through a corner-protonated intermediate or else predominate retention of configuration would not be found in most protonation reactions. An example is the ring opening of 1-methyl-2-phenylcyclopropanol (4). If a corner substituted intermediate were formed, attack to give 49 might be expected to be favored since it involves attack from the less hindered direction. Yet in this and in other cases complete retention occurs. Does this retention occur *via* an edge-protonated cyclopropane intermediate as

---

a) It has been demonstrated that electrophilic centers are not stabilized by the face of a cyclopropane ring [69,70]

in Mechanism II? This in unlikely because of the relative insensitivity of the direction of ring opening to substituents at the carbon to which the proton becomes attached. For instance attack of a proton on *4* gives rise to nearly equal amounts of two isomeric ketones *51* and *52*. If edge-protonated intermediates were involved it is hard to see why *53* would

not be greatly favored over *54* since the positive charge in *53* can be stabilized by the phenyl group. It seems more likely that proton addition in this case occurs through a single-step reaction, Mechanism I, and that this reaction occurs with complete retention of configuration. Assuming that C—H bond making balances C—C bond breaking, little or no charge build up need occur on the carbon undergoing electrophilic attack, and so little or no substituent effect would be observed.

A great deal of sense can be made of the experimental data if we assume that there is a competition in any cyclopropane protonation reaction between corner protonation and direct one-step opening with the latter pathway the more favored the greater the driving force for ring-opening. We therefore expect the one-step mechanism to predominate for *4*, and for other systems in which a good carbonium ion is formed. When the carbonium ion which would be formed is not especially stable (a primary or a secondary cation) corner protonation may become competitive with direct opening. This appears to be the case in the tricyclic system *14* in which inversion is observed on reaction with D$^+$. (See Eq. (9)). If we assume, as La Londe [9] did, that attack at the internal cyclopropane bond is difficult because of steric hinderance, then two

corner-substituted cyclopropanes, *55* and *56*, can be formed. Products which might be assumed to arise from both are formed in this reaction, but the product of stereochemical interest (*57*) would be that expected from *55* and would show *inversion* of configuration by deuterium.

Can all of the present results be accommodated without the need of postulating an edge-protonated cyclopropane ring? The tracer work of Baird [17], Lee [32], and Deno [31] can be fitted to a scheme in which corner protonation and direct opening with 1,3-hydride shifts are competitive, but an edge protonated cyclopropane ring accommodates the data in a much simpler fashion [b]. An edge protonated cyclopropane

---

[b] However, Deno [27] has questioned the validity of the non-statistical deuterium scrambling results.

is also implicated in the results from cyclopropane acylation [55]. If only corner substituted species were important, then structure *58* should be of lower energy than *59*, since the positive charge in *59* is shared by a carbon adjacent to the carbonyl group. But structure *58* does not lead to the correct product *47* upon attack by chloride ion. On the other hand edge protonated structure *60* should be more stable than *61* and also leads to the correct product. Although product structures do not always mirror the relative stability of intermediates, they do suggest that in this case edge-protonated intermediates may be involved.

B. Mercuration

In contrast to protonation, mercuration of a cyclopropane ring is strongly influenced by steric effects, so much so that they determine the overall stereochemistry of the ring opening. It seems quite clear that mercury attacks a cyclopropane ring in such a way that substituents at both ends of one of the edges can affect the energy of the transition state. At first sight, then, an edge mercurated cyclopropane (*63*) appears to be a reasonable intermediate.

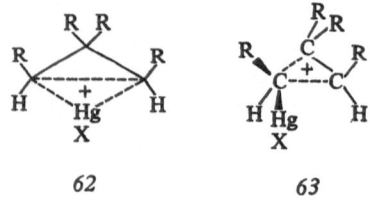

Several pieces of evidence argue against such an intermediate. The first is the large ϱ value reported [45] for the reaction of phenylcyclopropane with mercuric acetate. Steric effects argue that attack of mercury should be on the C2—C3 bond. An edge-mercurated intermediate (*62*) formed in the rate determining step would not give a role to charge stabilization by a substituent at C1. It has also been reported [22] that 1,2,2-trimethylcyclopropanol (*3*) reacts at the same rate with mercuric acetate as does 1-methylcyclopropanol. It seems, therefore, that a methyl group at each end of the C—C bond has a greater effect than two methyl groups at one end and none at the other. Finally, the stereochemical results of mercuration of *cis*-1,2,3-trimethylcyclopropane are much easier to accommodate if ring-opening occurs from a corner-substituted cyclopropane (*63*) than from an edge-substituted structure (*62*).

cis-1,2,3-Trimethylcyclopropane on mercuration gives 68% inversion by mercury and 32% retention. If an edge-mercurated cyclopropane were the sole intermediate, ring-opening with retention is easy to visualize, but opening with inversion requires a large atomic reorganization, with extensive movement of mercury. It is hard to see why these two fundamentally different processes should have nearly the same activation energy. Ring-Opening from a corner-mercurated intermediate (*64*) should occur from the rear of the non-mercurated carbons, C2 and

64

65

inv-inv

66

ret-inv

C3. Attack from the rear at C2 (path *A*) gives *65* in which both electrophile and nucleophile have reacted with inversion. If path *B* is followed, (attack at the rear of C3), the nucleophile again enters with inversion while the carbon mercury bond is formed with retention of configuration. According to this picture, the stereochemistry of the electrophilic attack in this system is really determined by the nucleophile. Since attack at the rear of C2 and C3 is nearly the same sterically it is not surprising that nearly equal amounts of *65* and *66* are formed in the reaction. Even if we allow free rotation of C1, the pentavalent mercurated carbon atom, the stereochemical results do not change. Such an intermediate (*64*) would place positive charge on C2 and C3 and could account for the large ϱ values observed. We therefore favor a corner-mercurated intermediate with the rate-determining step for the overall reaction leading to this intermediate. This picture is discussed in orbital terms in the following section.

C. Walsh Orbitals as Aids in the Interpretation of Electrophilic Attack on Cyclopropanes

The Walsh [71] orbitals for cyclopropane are particularly useful in visualizing the possible pathways for electrophilic attack on cyclopropane. The three bonding Walsh orbitals are given below [72]. Attack by an elec-

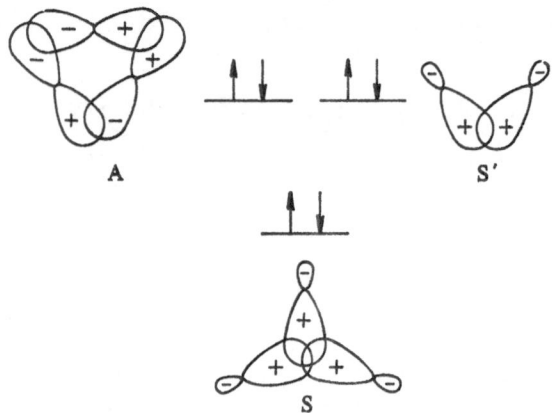

trophile on the higher energy S' orbital would lead to the best overlap, transforming it into a C-Electrophile bond. The A Walsh orbital would lead to the new C—C bond, and the low energy three-center S bond would remain essentially unchanged, giving rise to the three-center bond in the product. The origin of the bonds in the product is indicated in (67).

67

Notice that in this picture attack to give a corner-substituted intermediate will occur along an edge and will be subjected to steric hindrance by substituents at both ends of the bond, especially if the bond is symmetrically substituted.

A second attractive way for an electrophile to attack a cyclopropane might be at the anti-bonding portion of orbital A. This would require attack by an anti-symmetric orbital of the electrophile; it could lead to an edge-substituted cyclopropane in which the edge-substituted carbon-carbon bond is actually strengthened by breaking of its antibonding component (68). Such an intermediate might be particularly important

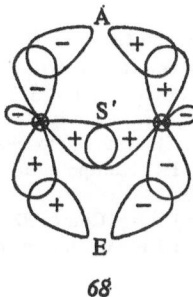

68

in reaction of cyclopropanes with transition metal ions.

VIII. Conclusions

Cyclopropanes can undergo attack by electrophiles with either inversion or retention of configuration at the carbon to which the electrophile becomes attached. The stereochemistry at this center, plus the fact that the nucleophile enters with inversion in nearly all cases, can best be accounted for by an intermediate, corner-substituted cyclopropane for many ring-opening reactions. In some cases a direct, single step process may compete with this mechanism. The possibility of an edge-substituted cyclopropane as a reaction intermediate under special circumstances cannot be ruled out at present.

References

1) Kramer, G. M.: J. Am. Chem. Soc. *92*, 4344 (1970).
2) DePuy, C. H., Arney, W. C., Jr., Gibson, D. H.: J. Am. Chem. Soc. *90*, 1830 (1968).
3) DePuy, C. H., Breitbiel, F. W., DeBruin, K. R.: J. Am. Chem. Soc. *88*, 3347 (1966).
4) Cristol, S. J., Lim, W. Y., Dahl, A. R.: J. Am. Chem. Soc. *92*, 4013 (1970).
5) Hendrickson, J. B., Boeckman, R. K., Jr.: J. Am. Chem. Soc. *91*, 3269 (1969).
6) Moore, W. R., Taylor, K. G., Müller, P., Hall, S. S., Gaibel, Z. L. F.: Tetrahedron Letters *1970*, 2365.
7) Nickon, A., Lambert, J. L., Williams, R. O., Werstiuk, N. H.: J. Am. Chem. Soc. *88*, 3354 (1966).

8) Wharton, P. S., Bair, T. I.: J. Org. Chem. *31*, 2480 (1966).
9) LaLonde, R. T., Ding, J-Y, Tobias, M. A.: J. Am. Chem. Soc. *89*, 6651 (1967).
10) Hogeveen, H., Roobeek, C. F., Volger, H. C.: Tetrahedron Letter *1972*, 221.
11) Warnet, R. J., Wheeler, D. M. S.: Chem. Commun. *1971*, 547.
12) Hammons, J. H., Probasco, E. K., Sanders, L. A., Whalen, E. J.: J. Org. Chem. *33*, 4493 (1968).
13) DePuy, C. H., Andrist, A. H., McGirk, R. H.: Unpublished results.
14) LaLonde, R. T., Debboli, A. D., Jr.,: J. Org. Chem. *35*, 2657 (1970).
15) Hendrickson, J. B., Boeckman, R. K., Jr.: J. Am. Chem. Soc. *93*, 4491 (1971).
16) Peterson, P. E., Thompson, G.: J. Org. Chem. *33*, 968 (1968).
17) Baird, R. L., Aboderin, A. A.: J. Am. Chem. Soc. *86*, 252 (1964).
18) McKinney, M. A., Smith, S. H., Hempelman, S., Gearen, M. M., Pearson, L.: Tetrahedron Letters *1971*, 3657.
19) McKinney, M. A., So, E. C.: J. Org. Chem. *37*, 2818 (1972).
20) de la Mare, P. B. D., Bolton, R.: Electrophilic additions to unsaturated systems, p. 26. New York: Elsevier 1966.
21) Parnes, Z. N., Khotimskaya, G. A., Kudryrotsev, R. V., Lukina, M. Yu., Kursanov, D. N.: Dokl. Akad. Nauk SSSR *184*, 615 (1969).
22) DeBoer, A., DePuy, C. H.: J. Am. Chem. Soc. *92*, 4008 (1970).
23) Clark, J.: Ph. D. Thesis, University of Colorado, 1968.
24) McManus, L. D., Rogers, N. A. J.: Tetrahedron Letters *1969*, 4735.
25) Pittman, C. U., McManus, S. P.: J. Am. Chem. Soc. *91*, 5915 (1969).
26) Cristol, S. J., Harrington, J. K., Morrill, T. C., Greenwald, B. E.: J. Org. Chem. *36*, 2773 (1971).
27) Deno, N. C., Billups, W. E., La Vietes, D., Scholl, P. C., Schneider, S.: J. Am. Chem. Soc. *92*, 3700 (1970).
28) Collins, C. J.: Chem. Rev. *69*, 543 (1969).
29) Lee, C. C.: Progress in physical organic chemistry, Vol. 7, p. 129 (1970).
30) Baird, R. L., Aboderin, A. A.: Tetrahedron Letters *1963*, 235.
31) Deno, N. C., La Vietes, D., Mockus, J., Scholl, P. C.: J. Am. Chem. Soc. *90*, 6457 (1968).
32) Lee, C. C., Gruber, L.: J. Am. Chem. Soc. *90*, 3775 (1968).
33) Lee, C. C., Chwang, W. K-Y., Wan, K-M.: J. Am. Chem. Soc. *90*, 3778 (1968) and earlier papers.
34) Aboderin, A. A., Baird, R. L.: J. Am. Chem. Soc. *86*, 2300 (1964).
35) Karabatsos, G. J., Orzech, C. E., Jr., Meyerson, S.: J. Am. Chem. *87*, 4394 (1965).
36) Olah, G. A., White, A. M., DeMember, J. R., Commeyras, A., Lui, C. Y.: J. Am. Chem. Soc. *92*, 4627 (1970).
37) McAdoo, D. J., McLafferty, F. W., Bente, P. F., III: J. Am. Chem. Soc. *94*, 2027 (1972).
38) Chong, S-L., Franklin, J. L.: J. Am. Chem. Soc. *94*, 6347 (1972).
39) Saunders, M., Vogel, P., Hagen, E. L., Rosenfeld, J.: Accounts Chem. Res. *6*, 53 (1973).
40) Olah, G. A., Baker, E. B., Evans, J. C., Tolgyesi, W. S., McIntyre, J. S., Bastien, I. J.: J. Am. Chem. Soc. *86*, 1360 (1964).
41) Levina, R. Ya., Gladshtein, M.: Dokl. Akad. Nauk SSSR *71*, 65 (1950).
42) Lukina, M. Yu.: Russ. Chem. Rev. *31*, 419 (1962).
43) Nesmeyanova, O. A., Lukina, M. Yu., Kazanskii, B. A.: Dokl. Akad. Nauk SSSR *153*, 114 (1963).
44) Nesmeyanova, O. A., Lukina, M. Yu., Kazanskii, B. A.: Dokl. Akad. Nauk SSSR *153*, 357 (1963).

45) Ouellette, R. J., Robins, R. D., South, A., Jr.: J. Am. Chem. Soc. *90*, 1619 (1968).
46) McGirk, R. H., DePuy, C. H.: J. Am. Chem. Soc., in press.
47) McGirk, R. H.: Ph. D. Thesis, University of Colorado, 1971.
48) Deno, N. C., Lincoln, D. N.: J. Am. Chem. Soc. *88*, 5357 (1966).
49) Cristol, S. J., LaLonde, R.T.: J. Am. Chem. Soc. *80*, 4355 (1958).
50) Lambert, J. B., Black, R. D. H., Shaw, J. H., Papay, J. J.: J. Org. Chem. *35*, 3214 (1970).
51) Arney, W. C., Jr.: Ph. D. Thesis, University of Colorado, 1969.
52) DePuy, C. H., Jones, H. L., Moore, W. M.: J. Am. Chem. Soc. *95*, 477 (1973).
53) Hart, H., Schlosberg, R. H.: J. Am. Chem. Soc. *88*, 5030 (1966).
54) Hart, H., Schlosberg, R. H.: J. Am. Chem. Soc. *90*, 5189 (1968).
55) Hart, H., Levitt, G.: J. Org. Chem. *24*, 1261 (1959).
56) South, A., Jr., Ouellette, R. J.: J. Am. Chem. Soc. *90*, 7064 (1968).
57) Ouellette, R. J., Williams, S.: J. Org. Chem. *35*, 3210 (1970).
58) Moon, S.: J. Org. Chem. *29*, 3456 (1964).
59) Ouellette, R. J., Miller, A., South, A., Jr., Robins, R. D.: J. Am. Chem. Soc. *91*, 971 (1969).
60) Paquette, L. A.: Accounts Chem. Res. *4*, 280 (1971).
61) Oullette, R. J., Levin, C.: J. Am. Chem. Soc. *90*, 6889 (1968).
62) Young, L. B.: Tetrahedron Letters *1968*, 5105.
63) Graham, W. A. G., Stone, F. G. A.: Chem. Ind. *1957*, 1096.
64) Rickborn, B., Wood, S. E.: J. Am. Chem. Soc. *93*, 3940 (1971).
65) Tipper, C. F. H., Walker, D. A.: Chem. and Ind. *1957*, 730.
66) Hixon, S. S., Garrett, D. W.: J. Am. Chem. Soc. *93*, 5294 (1971).
67) Cacace, F., Guarino, A., Speranza, M.: J. Am. Chem. Soc. *93*, 1088 (1971).
68) Incremona, J. H., Upton, C. J.: J. Am. Chem. Soc. *94*, 301 (1972); Maynes, G. G., Applequist, D. E.: J. Am. Chem. Soc. *95*, 856 (1973).
69) Sherrod, S. A., Bergman, R. G., Gleicher, G. J., Morris, D. G.: J. Am. Chem. Soc. *92*, 3469 (1970).
70) Radom, L., Pople, J. A., Buss, V., Schleyer, P. v. R.: J. Am. Chem. Soc. *93*, 1813 (1971).
71) Walsh, A. D.: Trans. Faraday Soc. *45*, 179 (1949).
72) Hoffmann, R.: J. Am. Chem. Soc. *90*, 1475 (1968).

Received November 27, 1972

Reactivity of Cycloalkenecarbenes

Prof. Dr. Heinz Dürr

Institut für Organische Chemie, Universität des Saarlandes, Saarbrücken

Contents

The field of carbenes has been intensively investigated for the last two decades[1]. These studies have shown that carbenes as well as nitrenes and arynes play an important role in organic chemistry as reactive intermediates[2]. Not only have these studies made it possible to understand many organic reactions in detail, they have also given use to a simple classification of organic reactions on the basis of these reactive intermediates.

In the last decade a new class of carbenes has been recognized: the Cycloalkenecarbenes. These can be defined as cyclic carbenes which possess a carbene C atom attached to a conjugated π system.

Within this definition, one can construct the following series of cycloalkenecarbenes:

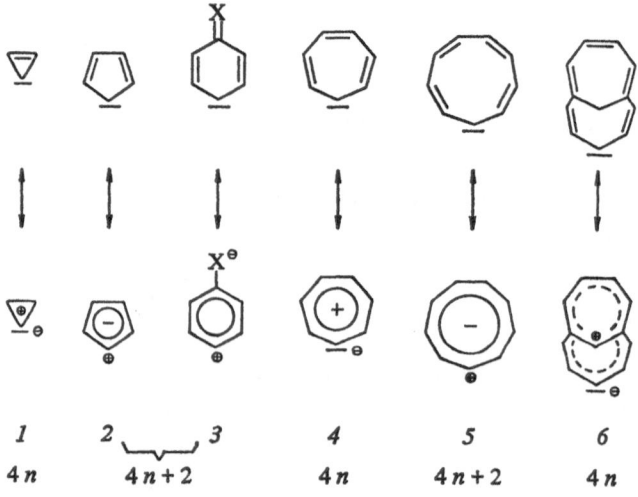

which contain a total of $4n$ or $4n+2$ electrons. (Note that σ electrons are not taken into account.) Resonance structures for these cycloalkenecarbenes — as shown above — have been proposed leading to two different types of cycloalkenecarbenes. Carbenes with 3-, 7- and 11-membered rings ($4n$ electrons) show a negative charge on the carbene center; those with 5-, 6- and 9-membered rings have a positive charge on the carbene center. According to the dipolar formula we may expect carbenes to have either a certain nucleophilic (*1, 4* and *6*) or a certain electrophilic (*2, 3, 5*) character. However one should realize that this resonance description is correct only for electrons being in the *same plane*, otherwise one is dealing with *two different electronic states* (see definition p 108).

The nomenclature of the cycloalkenecarbenes follows the same rules as apply for the other carbenes. The correct name is obtained by attaching either the prefix carbena- or the ending -ylidene to the cycloalkene skeleton. The 3-membered ring carbene can thus be called carbena-cyclopropene or cyclopropenylidene. Both terms will be used in this review, which will:

1. discuss the electronic structure of the cycloalkenecarbenes, and

2. describe the reactivity, the potential nucleophilicity or electrophili-city, and the chemical multiplicity of these carbenes.

The main emphasis is given to investigations where both these aspects have been studied. This condition excludes many excellent reports of preparative work on cycloalkenecarbenes [3]. Since the field of cycloalkenecarbenes still poses many problems, this report is an attempt to summarize the results obtained to date and to define the problems which still have to be explained.

I. Electronic Structure of Cycloalkenecarbenes – Extended HMO Calculations

In simple carbenes (or nitrenes) the energy levels of the carbene orbitals all possess the same energy, *i.e.* they are degenerate.

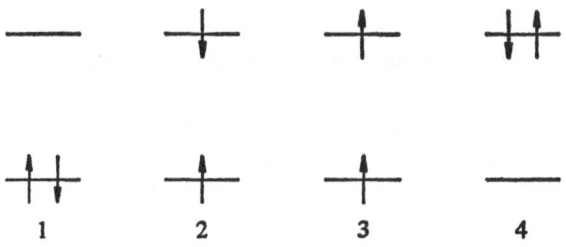

This is the reason why each of these orbitals is occupied by only one electron, the resulting state being a triplet ground state [1]. Certain factors can destroy this degeneracy and it is possible to have a carbene whose ground state is a stabilized singlet. If we consider two carbene orbitals of different energy, then there are four possible configurations for the two carbene electrons [4]:

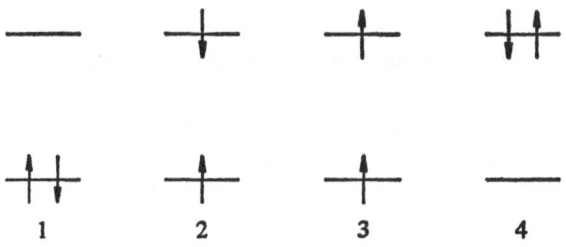

Configuration 1 is the ground configuration, 2 and 3 are singly excited, and 4 is doubly excited. Electron interaction places 2 at higher energy than the triplet 3 (Hund's rule). Configuration interaction mixes 1 and 4.

According to Hoffmann [4] an empirical rule[5] shows that the level splitting of the p_y and p_x orbitals must be of the order of $\Delta E = 1.5\text{--}2.0$ eV. Thus a singlet ground state is stable, *i.e.* both electrons are in the lower energy level. If the level splitting is less than 1.5 eV, then the ground state will be of triplet multiplicity.

The degeneracy of the p_x and p_y ($\equiv \sigma$) orbitals can be destroyed by two effects:

1. Reduction of the R—C—R angle θ from its normal value of 180 °. This reduction does not affect p_x; $p_y \equiv \sigma$ acquires s character and is therefore stabilized (approaches an sp^2- hybrid).

2. Conjugation with π systems having either unoccupied π or occupied π orbitals of correct symmetry. This effect can also operate in cycloalkenecarbenes, to break the degeneracy of the carbene orbitals.

We can distinguish two cases where the carbene interacts with π systems:

4n-Carbenes: a $4n + 2\,\pi$ system with a carbene C gives a $4n$ carbene.

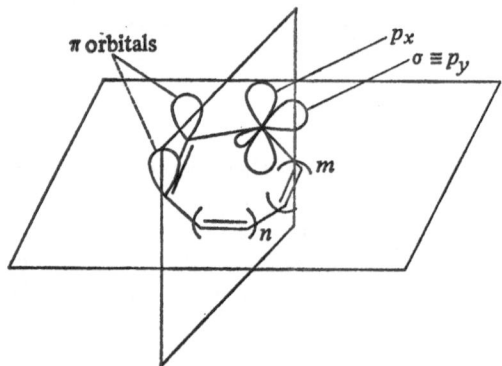

In cyclopropenylidene ($m=n=0$) the highest occupied polyene radical (S) has the correct symmetry for interaction with the p_x orbital (S). Mixing of these orbitals leads to two new orbitals of different energy: p_x is destabilized, and σ is stabilized by bending, which reinforces the splitting of the σ and the delocalized p level. This is shown in the interaction diagram for cyclopropenylidene (*1*).

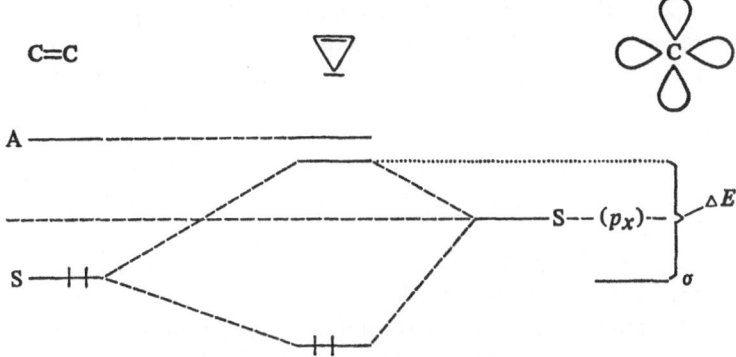

The energy difference between p_x (delocalized) and σ is $\Delta E = 3.17$ eV. Therefore the two carbene electrons are most favorably placed in the σ orbital so that we find a singlet ground state for cyclopropenylidene [4]. This result of extended HMO calculations[4] has not yet been proven experimentally. From these calculations a general rule can be derived: the interaction of carbene orbitals with a $4n + 2$ system should always lead to a large splitting of σ and p_x, giving more or less stable singlet ground states.

4n + 2 Carbenes: Conjugation of the carbene center with a polyene system of $4n$ π electrons means that the p_x orbital and the lowest unoccupied orbital of the π system must have the correct symmetry for interaction. This interaction again gives rise to the stabilization of p_x (delocalized), but the level splitting in this case is much smaller, since p_y has also been stabilized by bending. The simplest case is that of cyclopentadienylidene ($m=0$, $n=1$) for which the interaction diagram is:

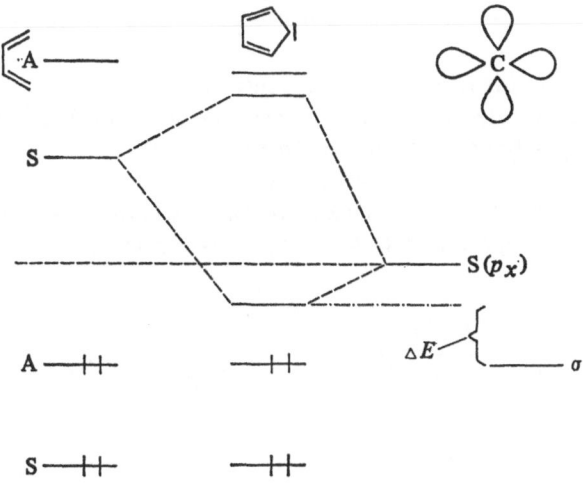

Cyclopentadienylidene (*2*) shows a very small splitting of the carbene energy levels, $\Delta E = 0.13$ eV. The energy of the new p_x and σ is very close, and therefore a triplet ground state for cyclopentadienylidene is expected. This is in fact borne out by ESR measurements [6] (see p. 109).

An extended HMO calculation for carbena-cyclohexadienone (*3*) [4b] gives slightly different results. The σ orbital of *3* is about 1.4 eV higher than the p_x orbital. This energy difference comes close to the ΔE value of 1.5—2.0 eV which is necessary for a singlet ground state. Therefore for *3* a triplet ground state is expected; this is confirmed by ESR measurements (see p. 109). But the singlet state should also be reached quite easily. The nucleophilic or electrophilic character of the cycloalkene-carbenes has also been evaluated by extended HMO calculations [4], and information can be gained from the net charge on the carbene center of the singlet species. The figures for this are shown in Table 1 together with the splitting of the energy levels.

From Table 1 we may conclude that the singlet carbenes *1, 4, 7a* and *7'* should be nucleophilic. Carbene *2* shows only a weak net negative charge for the σ^2 form and a slightly positive one for the p^2 configuration. Thus no great electrophilicity or nucleophilicity is to be expected for either the σ^2 or the p^2 carbene (see p.125). For greater clearness, the orbital schemes of the σ^2, p^2 and σp carbenes of cyclopentadienylidene are reproduced.

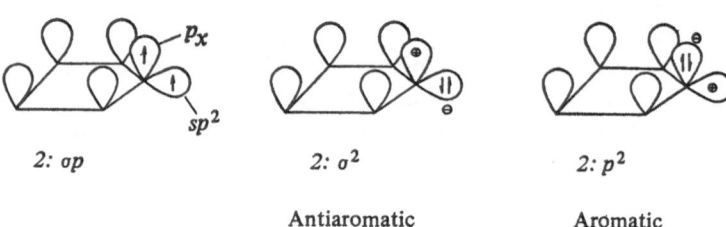

2: σp   *2: σ²*   *2: p²*

Antiaromatic   Aromatic

One important feature of cycloalkenecarbenes should be stressed here:

The resonance description *e.g.* delocalization for cycloalkenecarbenes *1—6* can be strictly applied only to the σ^2-carbenes of the $4n$- (*1, 4, 6*) and the p^2-carbenes of the $4n + 2$-series (*2, 3, 5*).

In *1, 4* and *6* the σ^2-carbene is aromatic whereas the p^2-carbene is antiaromatic. In *2, 3* and *5* this behaviour is reversed.

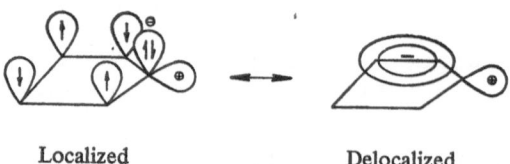

Localized   Delocalized

The 3 configurations σ^2, p^2 and σp of cycloalkenecarbenes are in fact species of different multiplicities which should show different reactivities. No studies to detect this differences have been carried out so far.

Table 1. Energy difference of the carbene levels and net charges on the carbene C atom [4]

Cycloalkene-carbene		Level splitting ΔE (in eV)	Net charge for configuration σ^2	p^2
		4n carbenes		
	(1)	3.17	−0.68	
	(7a)	3.45	−0.43	
	(4)	1.00	−0.86	
	(7')	1.77	−0.65	
		4n + 2-carbenes		
	(2)	0.13	−0.32[1]	+0.16[1]
	(3)	1.4	0.43 [4b]	

[1] Private Communication of Prof. R. Hoffmann.

II. ESR and UV Spectra of Cycloalkenecarbenes

A. ESR Spectra

The predictions of the HMO calculations can be checked by ESR spectroscopy. It is well known that the magnetic properties of radicals or triplets differ from those of molecules (closed shell) or singlets [1]. The

Table 2. ESR spectra of cycloalkenecarbenes having a triplet ground state

Cycloalkenecarbene	D (cm^{-1})	E (cm^{-1})	Angle at C$_1$	Reference
2 a	0.4089	0.012	—	6)
2 b	0.3777	0.0160	—	6)
2 c	0.4078 0.4084 0.4092	0.0283 0.0271 0.0283	135 °	6—8) 9,10) 11)
2 d	0.3991	0.0279	135 °	9,10)
3 a	0.3179	0.0055		12)
3 b	0.3284	0.0086		12)
3 c	0.3470	0.0010		12)
3 d	0.3333	0.0112		12)

Table 2 (continued)

Cycloalkenecarbene	D (cm^{-1})	E (cm^{-1})	Angle at C_1	Reference
3 e	0.04	0.002		13)
4 a : R = H	0.3787	0.0162	$\sim$150 °	10)
4 b : R = benzo	0.4216	0.0195		10)
8	0.4050	0.019	$\sim$150 °	9, 10)
7 b	0.52	—	—	14)
CN—C̈—H	0.8629	0	180 °	15)
HC≡C—C̈H	0.6276	0	180 °	15)

paramagnetic triplets give rise to ESR signals whereas the diamagnetic singlets do not, and thus one can easily distinguish between a singlet and a triplet state of a carbene. ESR spectra are usually obtained by photolyzing the corresponding diazocycloalkenes in a matrix at 77 ° or 4 °K, [6)] therefore the ground state of a carbene is normally observed under these conditions. Two parameters are derived from the ESR spectra D, which corresponds to the average separation of the two electrons, *i.e.* the degree of delocalization, and E, which is a measure of the deviation from a linear structure. In Table 2 the D and E parameters are given for the cycloalkenecarbenes studied thus far.

On the basis of the D and E values from Table 2, the cycloalkenecarbenes may be arranged in order of:

1. Decreasing D value:

   $7b > 4b > 2c \sim 2a > 8 > 2d > 4a > 2b > 3d > 3b > 3a >$ and

2. Decreasing E value:

   $2c \sim 1c > 4b \sim 8 > 4a > 2b > 2a > 3d > 3b > 3a$

The decreasing D values show — at least roughly — that electron-attracting groups delocalize the electron in the p_x orbital. The cyclohexadienonecarbenes (3) therefore show the smallest D parameters. Fusion to an aromatic ring seems to interfere with the delocalization of a cycloalkenecarbene, since the π electrons are more localized in the benzene system. The relatively large D value for 4b (0.4216 cm^{-1}) is attributed to steric interactions [10]. The order $2c > 2a > 2d > 2b$ parallels roughly the decreasing p_{Ka}-values, which are a measure of carbanionic stabilization [11].

If the E parameters corresponded to the internuclear angle at C-1, the order of decrease we would expect to be $4 > 3 > 2$: however, the actual sequence is: $4 > 2 > 3$. This has been explained by the different effects of the x and y components of the spin densities on the σ orbital at C-1 [6, 12]. From the $|E|/|D|$ value it was concluded that all cycloalkenecarbenes have angles which are large compared to the internuclear angle at C-1; thus all the bonds at C-1 should be bent, i.e. C-1 should not be in the ring plane [6, 10, 12]. This suggests that 4 is almost planar [10], allowing delocalization of the p_x electron. The electronic state of 1 has not been studied yet.

B. UV Spectra of Cycloalkenecarbenes

Very little work has been published in the area of the UV spectroscopy of cycloalkenecarbenes. The few investigations that have been carried out were done with the aim of finding out whether the electronic state at 77 °K is identical with that at room temperature. The results published so far are reproduced in Table 3.

Table 3. UV spectra of cycloalkenecarbenes

Cycloalkenecarbene	λ max (nm) at 77 °K	Reference
8	468	16, 17)
	300 ⎫ 465 ⎭	18)
	486[1])	19)

112

Table 3 (continued)

Cycloalkenecarbene	λ max (nm) at 77 °K		Reference
4a	383[1])	10 μ sec[2])	20, 21)
	395		
	486		

[1]) Identical with spectra obtained at room temperature by flash photolysis.
[2]) Lifetime.

As compared to *8*, *4a'* and *4a* show a bathochromic shift. This is obviously due to a larger delocalization of the p_x electron. There is practically no difference between the UV spectra of *4a'* and *4a*. The flash photolysis experiments demonstrate clearly that the spectroscopic information obtained at 77 °K also applies at ordinary temperature [19—21]). The lifetime of carbene *4a* at ambient temperature has been shown to be of the order of 10 μ sec [21]).

III. Generation of Cycloalkenecarbenes

Cycloalkenecarbenes are reactive intermediates and are not stable under normal reaction conditions, *i.e.* they react instantaneously to give stable products. This section reviews the current methods of generating cyclo-alkenecarbenes. Their reactions are thereafter described without regard to the method of preparation.

The simplest way of generating a carbene is by photolysis or pyrolysis of a suitable diazo compound. This reaction gives a free carbene species [1]) and is therefore the method most often used.

The cyclopropenylidene *1* and the heterocyclic carbenes *7* and *7'* have been prepared by different routes, thus leading most probably to car-benoid species since K-t-butylate is present in the system.

Cyclopropenylidene (1). The generation of *1* posed — and still poses — a serious problem. All the classical methods have been tried to synthesize

113

a suitable diazo precursor and have failed, or only slight amounts of
1 [22-24] have been generated. The only useful method for the generation
of *1* is thermolysis of the N-nitroso-carbamate *9* in the presence of a strong
base, leading to *1a* [25] via the intermediate *10:*

Cyclopentadienylidene (2). *2* was very easily obtained by photolysis
or thermolysis of the corresponding diazo-cyclopentadienes (*12*). These
were prepared by straightforward procedures (diazo-group transfer,
dehydrogenation of hydrazones, or Bamford-Stevens reaction of tosyl-
hydrazones) [26-30] from either cyclopentadienes (*11*) or substituted
cyclopentadienones (*13*).

Carbena-cyclohexadienone (3). Photolysis or thermolysis of *p*-quinone
diazides (*15*) easily led to *3*. The precursors of *3* are either *p*-aminophenols
(*14*), *p*-quinones (*17*) or anthrones (*16*). Diazotization of *14*, Bamford-
Stevens reaction of the tosyl-hydrazone of *17*, or diazo-group transfer to
16 afforded the corresponding *p*-quinonediazides *15* [31-34].

Cycloheptatrienylidene (*4*). Unsubstituted *4* could be obtained by thermal or photochemical decomposition of tropone tosylhydrazone salt (*19c*) [35, 36]. In the case of the dibenzo (*19a*) or tribenzo derivatives (*19b*), the diazocycloheptatrienes (*20a, b*) could be isolated [37]. Then photolysis of *20a, b* gave *4* in these cases too. An interesting entry in the cycloheptatrienylidene series consists of flash pyrolysis of phenyldiazomethane (*21*) followed by a rearrangement to *4* [38].

The 11-membered ring carbene was generated by thermolysis recently. To ensure coplanarity a brigded precursor such as 4.9-methano (11) annulenone-tosylhydrazone was used to give thus 4.9-methano (11) annulenylidene 6 [39]. The isomeric annulenylidene 6' was synthesized from the tosylhydrazone of 1.6-methano-annulene-2-carboxaldehyde [39].

Hetero-cycloalkene-carbenes (7c, 7', 21). These carbenes could — until now — only be generated as carbenoid species. So treatment of 1.3-diphenylimidazolium halide (24) with alkoxide gave 7c [40], whereas N-alkylbenzthiazolium halides (25) in the presence of t-amines yielded 26 [41]. Pyrolysis of the zwitterions 27 led to the formation of 7' as intermediates [42].

R[1], R[2] = H, *o*-phenylene

IV. Reactivity of Cycloalkenecarbenes

A. Relative Rate of Insertion Reactions

Cycloalkenecarbenes — like other carbenes — insert readily in an intermolecular reaction in the C—H bonds of alkanes. This reaction can

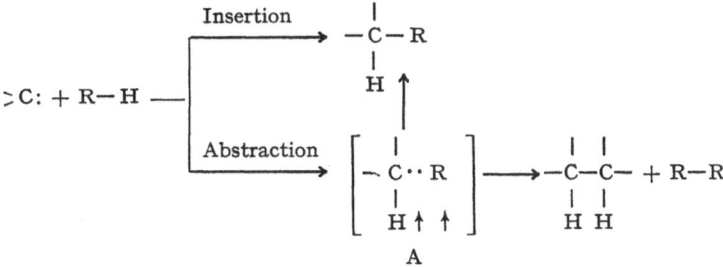

proceed by two different modes, either by concerted insertion or by a two step abstraction process [1].

The insertion reaction gives only a single product whereas the abstraction process leads primarily to a radical pair A which can then combine to form three different compounds. A singlet carbene usually gives insertion while a triplet carbene reacts via abstraction. These two processes can easily be distinguished in gas-phase reactions. In the condensed phase, however, the radical pair A is trapped in the solvent cage. Recombination of the radicals then leads to a single product which is identical with the insertion product. In special cases abstraction reactions can be observed in the condensed phase too. The photolysis of diazofluorene (12c) in cyclohexane gives the products 9-cyclohexyl-fluorene (28) and 9.9'-difluorenyl (29), formed by an abstraction reaction [43].

Another clear-cut example of abstraction was observed in the reactions of anthronylidene (3f) [33, 34] with cyclohexane or toluene. All possible dimerization products 30, 31 and 32 can be isolated in this case. These abstraction reactions are attributed to triplet carbenes 2c and 3f. All other cycloalkenecarbenes show normal C—H insertion reactions (for the exceptional behaviour of di- and tribenzocycloheptatrienylidene see p. 137).

Photolysis of phenyl-substituted diazocyclopentadienes (12) in cyclohexane gives excellent yields of substituted cyclohexylcyclopentadienes (33) [27, 44].

Use of alkanes with p, s or t C—H bonds will give three different insertion products. The ratios of which yield the relative rate of insertion. Carbena-cyclopentadiene (2a) generated photochemically inserts in the C—H bonds of 2.3-dimethylbutane to give two products 34 and 35 in the ratio 1.22:1 [26].

Corrected for statistical factors, insertion occurs 7.32 times faster with the tertiary than with the primary C—H bonds of 2.3-dimethylbutane.

Similiar experiments have been carried out with 2.5-diphenyl-cyclopentadienylidene (2h) [45]. These are the only data so far available on the relative insertion rates of cycloalkenecarbenes. Table 4 shows a

117

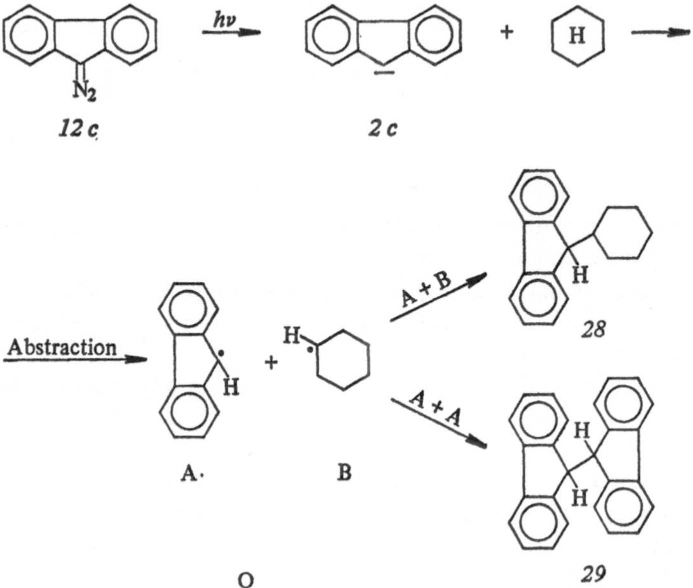
hν
H
12c
2c
Abstraction
H
H
A·
B
A+B
A+A
H
28
H
H
29

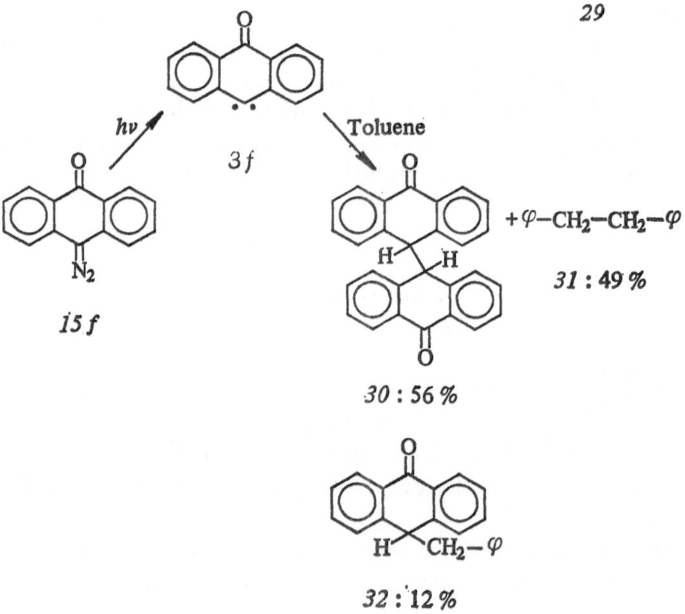
hν
Toluene
3f
15f
+φ–CH2–CH2–φ
31 : 49 %
30 : 56 %
H CH2–φ
32 : 12 %

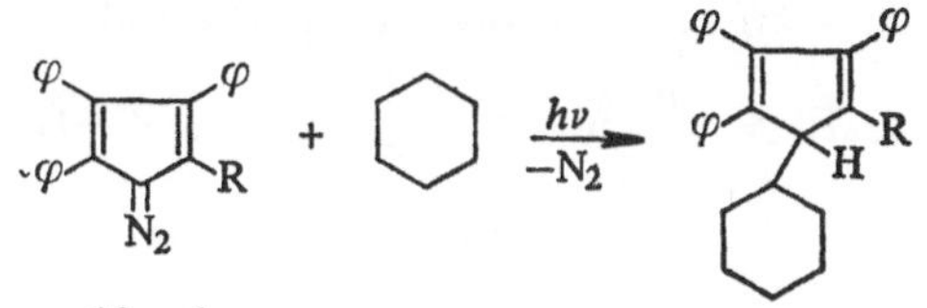

12 e, f   *33*

$R = H, \varphi$

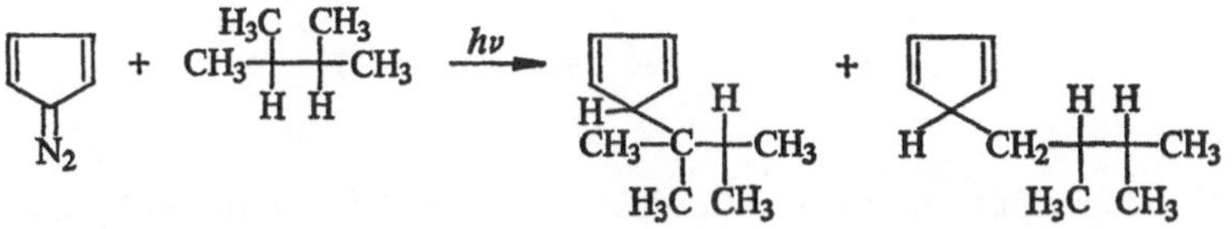

12 a   *34*   *35*

comparison of relative rates of *2a* and *2h* with the results obtained for simple carbenes.

Table 4. Competition constants for insertion of cycloalkenecarbenes and simple carbenes

Precursor	Mode of generation	Alkane	Insertion rate		Ref.
			$2°/1°$	$3°/1°$	
 *12 a*	*hv*	2.3-Dimethyl-butane	—	7.3	26)
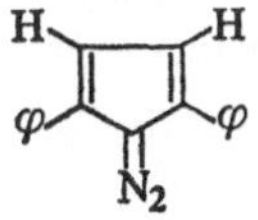 *12 h*	*hv*	2.3-Dimethyl-butane	—	4.5	45)
CH_2N_2	*hv*	2.3-Dimethyl-butane	—	1.2	46)
CHN_2CO_2R	*hv*	2.3-Dimethyl-butane	—	2.9	47)
$\varphi-CHN_2$	*hv*	Pentane	8.3	—	48)
$CN_2(CO_2R)_2$	*hv*	2.3-Dimethyl-butane	—	12.5	46)
$C(CN)_2N_2$	Δ	Isobutane	—	12.0	49)

From these data an order of decreasing reactivity can be derived:

$$:CH_2 > H\ddot{C}-CO_2R > \text{[cyclopropenylidene]} > \varphi-\text{[cycloheptatrienylidene]}-\varphi > \varphi-C-\ddot{H} > :C(CO_2R)_2$$

$$\approx :C(CN)_2$$

This decreasing reactivity can be explained as follows:

a) The contribution of the canonical structure A is increased by carbene substituents R with electron-attracting abilities. This is equivalent to a stabilization of the transition state of the reaction. The greater this stabilization, the greater the discrimination of a certain carbene, *i.e.* the discrimination increases in the above sequence from left to right.

b) The more stable the carbonium ion in the transition state A, the more selective is a carbene. This explains the greater insertion rate in the order $3° > 2° > 1°$ C—H bonds.

Relative rates of insertion have not been measured for *5. 1* and *4* do not insert at all[a].

B. Relative Rate of Addition Reactions versus C—H Insertions

Cycloalkenecarbenes react with olefins either by [1 + 2] cycloaddition to the double bond or by insertion in the C—H bonds. Thus spirocyclopro-

---

[a] Private communication from Prof. W. M. Jones, University of Florida, Gainesville/USA.

panes (A) — which are of synthetic interest — or substituted alkene derivatives (B) are formed.

The relative rate of addition versus insertion can be used as a measure of carbene reactivity, too. For instance carbena-cyclopentadiene reacts with tetramethylethylene to give the spirocyclopropane *36* (A) and the alkenylcyclopentadiene *37* (I) [26], A: addition, I: insertion.

12 a	*36*	*37*

The ratio of *36*:*37* being 1.74. If this figure is corrected for statistical factors (12 equivalent C—H bonds) the A/I rate is 20.9. In the same manner the A/I rate has been determined for several other cycloalkenecarbenes; the results are presented in Table 5.

Table 5. Relative rate of addition versus insertion of cycloalkenecarbenes

Diazo compound	Alkene	A/I (corrected for statistical factor)	Ref.
12a, hv	2.3-Dimethyl-but-2-ene	20.9	26)
12e, hv	2-Methyl-but-2-ene	40	44)
12e, hv	trans-4-Methyl-pent-2-ene	34	44)
12e, hv	cis-4-Methyl-pent-2-ene	228[1]	44)

Table 5 (continued)

Diazo compound	Alkene	$A/_I$ (corrected for statistical factor)	Ref.
12 f	*cis*-4-Methyl-pent-2-ene	45	27)
12 g	2-Methyl-but-2-ene	≫45	28)
12 e	Cycloalkene	14.8	27)
		17.0	
		13.8	
		17.1	
12f, hv		72	44)
		20	
12f, hv		12	44)
		6	

Table 5 (continued)

Diazo compound	Alkene	$^A/_I$ (corrected for statistical factor)		Ref.
15 a	2-Methyl-but-1-ene	25		50)
H_2CN_2, $h\nu$	isobutene Addition/vinyl + allyl – insertion	16.2[2]	5.7[3]	51–53)
	2.3-Dimethyl-but-2-ene		8.3[3]	51, 52)
	Polyenes			
12e, $h\nu$	Cycloheptatriene	2.0		54)
12 h, $h\nu$	Cycloheptatriene	0.93		54)
12 f, $h\nu$	Benzene	21		55)

Table 5 (continued)

Diazo compound	Alkene	A/$_I$ (corrected for statistical factor)	Ref.
12g, hν	 Benzene	2.0	55)

1) This value is much too high.
2) Gas phase.
3) Condensed phase.

From Table 5 a rough order of reactivity for the cycloalkenecarbenes can be deduced:

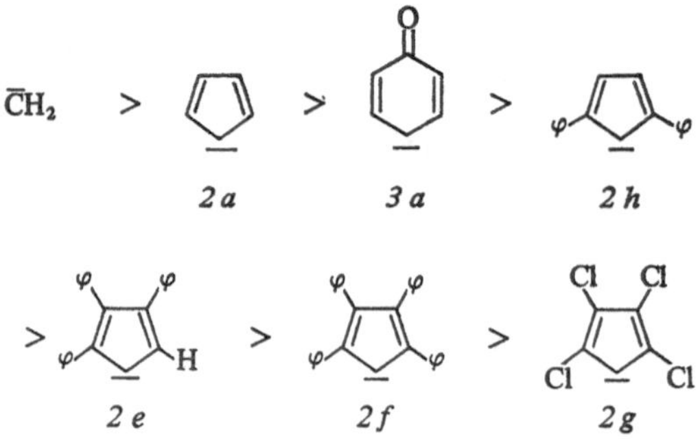

The rate of addition to a double bond increases in the sequence shown above, i.e. the cycloalkenecarbenes are less reactive (more stable) from left to right. The position of 2g is not reliable since no insertion with simple olefins has been reported. Again the pK_a-values would also be a measure of inductive and resonance effects. The A/I rate with polyenes is modified in comparison with the A/I values for simple olefins.

The decreasing A/I rate of 2e,f with cycloalkenes has been attributed to increasing steric hindrance in going from a 5-membered to an 8-membered ring. Steric effects mask the reactivity of the cycloalkenecarbenes in these reactions. In the case of the more stable (less reactive) cycloalkene carbenes the transition state is more product-like 56), i.e. the carbenes are more selective. Steric effects can obscure this pattern.

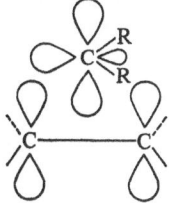 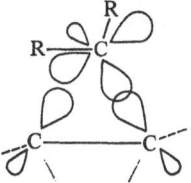

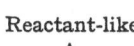

Reactant-like
A

Product-like
B

Transition state for addition

Highly reactive   Less reactive carbenes

C. Nucleophilicity or Electrophilicity of Cycloalkenecarbenes

Now that we have established — at least for some cycloalkenecarbenes — an order of reactivity, one important problem is still unresolved. Do the cycloalkenecarbenes *1—6* show some nucleophilic or electrophilic character as postulated (see p. 104)? This question can be tackled by three different approaches:

1. Competition experiments with olefin pairs of different nucleophilicity,
2. measuring the different rates and positions of substitution with benzene derivatives and
3. determining the product ratio in the reaction with azides.

1. Competition Experiments with Olefin Pairs

a) This method has been widely used to prove the electrophilic character of carbenes, but unfortunately, with the highly reactive carbenes, this procedure is a poor criterion for electrophilicity [57]. Nevertheless, the method has been applied to *2a* in the cycloalkenecarbene series. The relative rates of addition obtained by competition experiments are collected in Table 6.

Relative rates of addition of carbena-cyclopentadiene with olefins show no electrophilic order. In contrast to *2a*, for phenyl- and phenylbromocarbene a clear increase of addition rate with more nucleophilic alkenes is observed. For cyclohexadienylidene (*3e*) an electrophilic character was demonstrated. *3e* is not a cycloalkenecarbene stabilized by resonance but a simple divinylcarbene. *3e* should have the same steric requirements as *2a*. Thus, if there is a difference between *2a* and *3e* it cannot be due to steric [56] but rather to electronic effects. This means that there must be a special effect operating in *2a*.

125

Table 6. Relative rates of cycloalkenecarbene addition to olefins (in solution, relative to 2-methyl-but-2-ene)

Source of cycloalkene-carbene	R-CH=CH$_2$ R $=$ R (R always in brackets)	R R $\backslash=\diagup$ R	R R $\backslash=\diagup$	(hexagon)	R R $>=<$ R H	R R $>-<$ R R	Ref.
(structure) N$_2$, $h\nu$ ·12 a	1.25 (n-butyl) 0.93 (t-butyl)	— —	— —	1.33 —	1.00 (CH$_3$) —	0.99 (CH$_3$) —	26)
(structure) N$_2$ 15 e,$h\nu$	0.24 (n-butyl)	0.21 (CH$_3$ and i-propyl)	0.19 (CH$_3$ and i-propyl)	—	1.00 (CH$_3$)	1.23 (CH$_3$)	13)
(structure) ,$h\nu$	0.23 (t-butyl)	—	—	—	1.00 (CH$_3$)	1.07 (CH$_3$)	58)
C$_6$H$_5$— —CH=N$_2$, $h\nu$	0.29 (CH$_3$—CH$_2$)	0.58 (CH$_3$)	1.01 (CH$_3$)	—	1.00 (CH$_3$)	—	59)
C$_6$H$_5$ N $\backslash$C$\diagup$ ‖ Br N, $h\nu$	—	0.08 (CH$_3$)	0.22 (CH$_3$)	—	1.00 (CH$_3$)	1.52 (CH$_3$)	60)

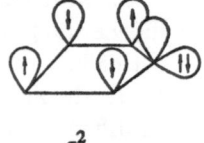

σ^2

Net charge at C$_1$: —0.32

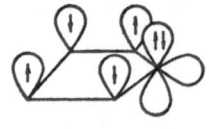

p^2

+0.16

The apparent lack of an electrophilic order for the addition of cyclopentadienylidene (see p. 126) may result from a product like transition state of a highly reactive carbene *2a*. Thus little charge would suggest a very high energy species with a very low activation for the addition. The reason could be a major contribution from the anti-aromatic σ^2-carbene of *2a* which should show some destabilisation in contrast to the aromatic p^2-carbene (see also [61]).

b) A very elegant approach for testing the electrophilic or nucleophilic character is the addition of a carbene to a styrene containing *p* substituents of differing electronegativity [62]. By this procedure relative addition rates are obtained which depending only on electronic and not on steric effects. This method has been used only very recently in carbene chemistry in a few cases [62]. From the reaction of carbenacycloheptatriene *4c* with differently *p*-substituted styrenes, relative rate constants were evaluated and are presented in Table 7.

A Hammett plot (Fig. 1) of the relative rates obtained for the *p* substituents: 4—OCH$_3$, 4—CH$_3$, H, 4—Cl, 4—Br and 3—Br, clearly shows that the addition rates are increased by electron-withdrawing groups.

From Fig. 1 a value of $\varrho = +1.05 \pm 0.05$ is derived, demonstrating definitely that *4c* is a nucleophilic species [62]. This beautiful result justifies the calculations and considerations described on p. 105. For comparison, the ϱ value for the electrophilic |CCl$_2$ is somewhat smaller but as expected, it shows the opposite sign ($\varrho = -0.61$) [62]. Unfortunately, no similiar data exist for *1*. There is, however, some hint that *1* also has nucleophilic properties: *1* does not add to electron-rich olefins but with electron-deficient olefins, such as dimethyl-fumarate, reacts very smoothly to afford the highly strained spirocyclopentenes *38* [22-24].

Table 7. Relative rate constants for the addition of cycloheptatrienylidene to substituted styrenes XC$_6$H$_4$CH $=$ CH$_2$

X in XC$_6$H$_4$CH $=$ CH$_2$	k_x/k_H[1]
4—OCH$_3$	0.51
4—CH$_3$	0.57
H	1.00
4—Cl	1.59[2]
4—Br	1.68
3—Br	2.22

[1] Average of two glpc determinations.
[2] nmr analysis of the mixture of the original adducts gave $k_{rel.} = 1.61$.

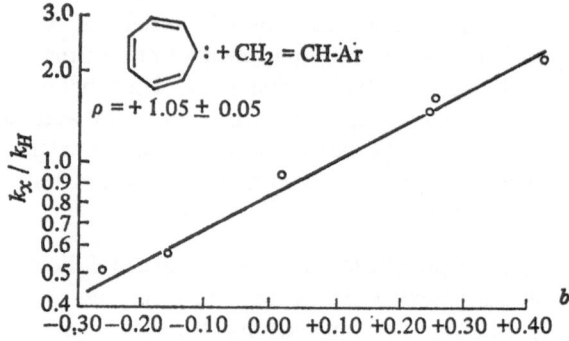

Fig. 1. $k_{rel.}$ vs. b values for the reaction of 2.4.6-cycloheptatrienone-p-toluene-sulfonylhydrazone sodium salt-derived cycloheptatrienylidene with substituted styrenes

1a   *38*

2. Different Rates and Position of Substitution with Benzene Derivatives

No quantitative data have been obtained by this procedure, but it demonstrates very clearly the character of the reacting species involved. It has been applied to 2.6-dichloro-carbena-cyclohexadienone (*3b*) [32)] but not to unsubstituted *3* itself. The results are nevertheless unambigous. Using this method, *3b* was reacted with a mixture of substituted

15b   *39*   *40*

benzenes and benzene and gave mixtures of o- and p-substituted 2.5-dichloro-4-hydroxy-biphenyls (39, 40) [32, 63]:

In some cases m-substituted biphenyls (41) are also obtained. From these experiments relative rate constants were measured for the overall substitution of benzene derivatives by $3b$. These data are collected in Table 8.

Table 8. Relative rate constants for overall substitution of benzene derivatives by $3b$ [32]

	Relative rates of substitution by		
(in $-C_6H_4X$)	$3b$	HNO_3/Ac_2O	φ ·
H	1	1	1
F	0.388	0.15	1.35
Cl	0.429	0.033	1.44
Br	0.345	0.03	1.75
CH_3O	1.276	—	2.5
CH_3CO_2	0.522	0.0037	—
CN	0.335	—	3.6

These relative rate constants are compared with the figures for nitration (HNO_3/Ac_2O) and phenylation (phenyl radicals) [32]. Table 9 gives partial rate factors for substitution with various benzene derivatives.

Table 9. Partial rate factors for substitution in C_6H_5X by $3b$

Benzene derivative	Partial rate factors for reaction with $3b$		
C_6H_5X	40	41	39
	(o)	(m)	(p)
F	0.72	0	0.88
Cl	0.85	0	0.88
Br	0.59	0	0.88
OCH_3	2.78	0	2.09
CO_2CH_3	0.55	0.49	0.37
CN	0.45	0.17	0.77

The data in Table 9 (no m-isomer) show that a radical mechanism can be ruled out. Table 8 shows that $3b$ is of even greater selectivity than the electrophilic nitration. These results clearly point to substitution by an electrophilic agent which can be formulated as $3b'$ [32].

This shows that the resonance representation of 3 (see p. 104) is of some importance and 3 is indeed an electrophilic intermediate. The

3 b $\longleftrightarrow$ 3 b'

resonance formula *3b'* gains weight by reason of the electron-attracting C=O group. However in the case of *3* the problem of either the σ^2- or the p^2-carbene being a major contributor arises again.

Table 10. Ratio of triaza and monoaza derivatives in the reaction of *26* with R'—N₃ (*42*) [64]

Azide [R'—N₃]⊕ BF₄⊖	Wavenumber of N≡N-frequency of [R'—N₃]⊕ BF₄⊖(cm⁻¹)	*43* (%)	*47* (%)	*43/47*
	2156[1])	72	< 1	ca. 72
	2155	73	< 1	ca. 73
	2158	24	11	62
	2158	18	30	0.6
p-CH₃-C₆ H₄ -SO₂ N₃	2137	< 1	56	ca. 0.02
	2130	—	47	—

[1]) Decreasing wavenumber is equivalent to decreasing electrophilicity.

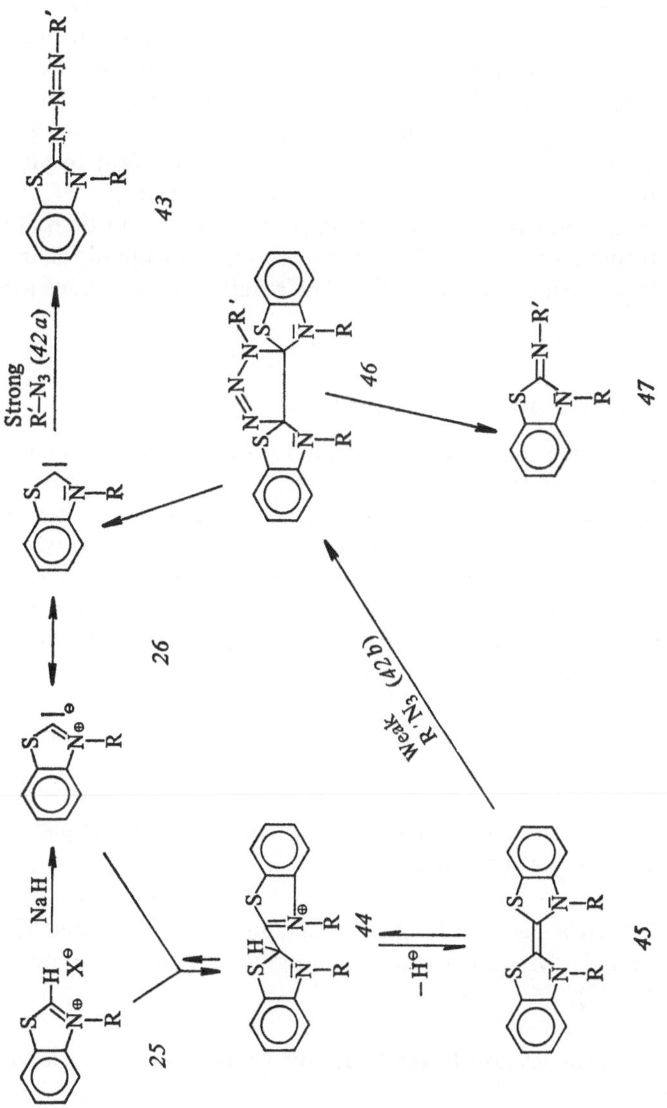

NaH
Strong
R'–N3 (42 a)
Weak
R'N3 (42 b)
–H⊕
25
26
43
44
45
46
47

3. Reaction with Azides of Differing Electrophilicity

This approach yields at least a rough estimate for the nucleophilicity of the reacting species. It has been used to test the nucleophilic character of benzthiazolene-ylidene *26* [40–42, 64]. The carbenoid *26* is generated from the N-alkylbenzthiazolium halide (*25*) in the presence of an azide of decreasing electrophilicity. With strongly electrophilic azides (*42a*), the carbenoid *26* reacts to a triazene *43*. Weakly electrophilic azides (*42b*) do not trap *26* immediately. The addition of *25* to *26* is faster, giving rise to the dimer *45* which then adds RN_3' (*42b*) to yield *46*. *46* can decompose further to *26* and the monoaza derivative *47*.

From the decreasing ratio *43/47* in the reaction with azides of decreasing electrophilicity we may deduce that *26* has nucleophilic character [64]. Table 10 gives the ratio of *43/47* in the reaction with various azides.

V. Chemical Multiplicity Studies

ERS studies have shown that the ground states of all cycloalkenecarbenes so far studied are of triplet multiplicity. A possible exception might be cyclopropenylidene *1*, whose ground state has not been determined yet (see p. 109). For cycloheptatrienylidene *4* only the dibenzo-derivative has been studied. The ESR investigations are usually carried out at low temperature. What however, is the spin state of the cycloalkenecarbenes at ordinary temperature? (See also p. 102). The solution to this problem is important since the delocalized formulae (see p. 108) are valid only for singlet states. The spin state of the reacting species of *1—6* can be elucidated by methods introduced by Skell [65], as summarized in the following statements:

1. Singlet carbenes add to olefins stereospecifically; triplet carbenes add non-stereospecifically.

2. Singlet carbenes should give predominantly C—H insertion with C—H bonds, triplet carbenes should give preponderantly abstraction (see p. 116).

3. Triplet carbenes add faster to 1.3-dienes than to mono-olefins.

Of these three rules, the first one has been most widely used and is the one mainly applied to cycloalkenecarbenes. The well-known [1 + 2] cycloaddition of carbenes to olefins is shown in the following scheme. All the steps for the concerted singlet cycloaddition are evident.

Singlet: *cis*-addition stereospecific   (a)

If in the triplet addition — which is a two-step reaction — rotation is faster than intersystem crossing and ring closure, *i.e.* $k_3 > k_1 \sim k_2$, then a mixture of *cis*- and *trans*-cyclopropanes is isolated. This rule is widely recognized though to date no sound theoretical basis has been found for it (a pyrazoline intermediate has been excluded in most cases by control experiments). If the cycloaddition follows path (a), it is called *stereospecific*. Carbenes having two different R groups can give rise to

both a syn- and an anti-addition product. If either one is preferred, the cycloaddition is called *stereoselective*. Only for unsymmetrically substituted cycloalkenecarbenes the latter type of reaction is possible. Stereospecifity can be increased by O_2 or NO, which act as triplet scavengers. It can be diminished by inert gases (N_2) or solvents (C_4F_8 or CH_2Cl_2) which favour intersystem crossing.

Table 11 presents chemical multiplicity studies for which criterion 1 (stereospecific addition) was almost exclusively used to demonstrate the spin state of the cycloalkenecarbenes.

133

From Table 11 the conclusion is obvious that the amount of non-stereospecificity is always larger in the $[1+2]$ cycloaddition of all cycloalkenecarbenes with *cis*- as compared with *trans*-olefins.

As can be seen from Table 11 all carbena-cyclopentadienes (*2*) add stereospecifically as singlets (S). The only exceptions are tetrachlorocyclopentadienylidene (*2g*) and the fluorenylidenes (*2c*, *2d*). The non-stereospecific addition — which is of the same order as for diphenyl-carbene [71] — in *2g* is most probably due to a heavy-atom effect of chlorine favouring intersystem crossing to the triplet (T). An aromatic ring as in *2c* and *2d* (as well as Br substituents) also facilitates the S—T interconversion step.

The reason for this fact is most probably due to the lowering of the energy of the lowest unoccupied MO (LUMO) in the order cyclopentadiene-, indenyl-, fluorenyl-anion[73]. A similiar effect should be operating in the cycloalkenecarbenes. Thus a stronger interaction with the p_x orbital (see p. 107) is possible resulting in a decreasing energy difference

Table 11. Chemical multiplicity studies of cycloalkenecarbenes

Carbene precursor (generation mode)	Stereospecificity		Abstraction	Ref.
	% *trans*-Cyclopropane from *cis*-olefin	% *cis*-Cyclopropane from *trans*-olefin		
12a (hv)	8[1] 2[2]	— 0		26) 66)
12e (hv)	< 1[2]	< 1[2]		44)
12f (hv)	< 1[2]	< 1[2]		27)
12g (hv)	10[1] 22[3] 58[4]	Trace[1] — —		28) 28) 45)

Table 11 (continued)

Carbene precursor (generation mode)	Stereospecificity		Abstraction	Ref.
	% *trans*-Cyclo-propane from *cis*-olefin	% *cis*-Cyclo-propane from *trans*-olefin		
12 c (hν) + C₆F₆	34[1] 45[2] 77[1] 65[4]	0[1] Trace 12[1] —	Little	67) 67) 9)
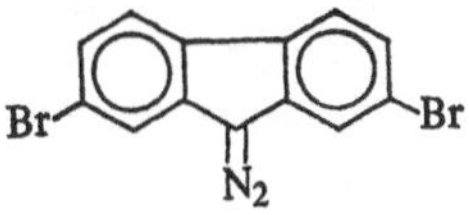 *12 d* (hν)	54[4]	—		9)
15 g (hν)	5[1] 33[1]	3[1]		68)
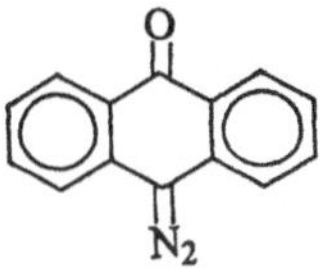 *15 f* (hν)	Not clear: photochemical *cis*-, *trans*-isomerization not excluded			34) 69)
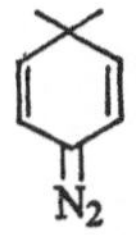 (hν)	8[2]	< 1[2]		13)
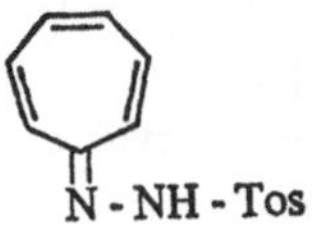 *20 c* (hν)	2[6]	2[6]		61a, b)
(hν + CH₂Br₂)	6[6]	2[6]		
(Δ) (thermal isomer-ization)	40[4]*	—		36)

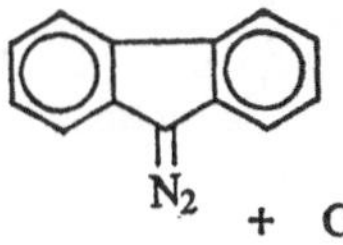

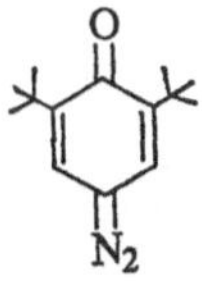

N - NH - Tos

135

Table 11 (continued)

Carbene precursor (generation mode)	Stereospecificity		Abstraction	Ref.
	% trans-Cyclopropane from cis-olefin	% cis-Cyclopropane from trans-olefin		
20a (hν)	0[1]	0[1]	23% dimer 2% abstraction	37)
20b (hν)	0[1]	0[2]	2% dimer	37)
CH_2N_2 (hν)	0[1]	0[1]		65)
φHCN_2 (hν)	3[1]	3[1]		70)
$\varphi_2 CN_2$ (hν)	34[1]	Trace[1]		71)
$(CN)_2 CN_2$ (Δ)	70[1]	30[1]		72)

[1]) cis- or trans-But-2-ene.
[2]) cis- or trans-4-Methyl-pent-2-ene.
[3]) cis- or trans-Pent-2-ene.
[4]) Dialkyl maleate or fumarate.
[5]) cis- or trans-1.2-Dichloroethylene.
[6]) cis- and trans-1.2-Dicyanoethylene.
*) See p. 137.

of the delocalized π and the σ orbital. As a consequence the triplet carbene becomes more favored in the order 2a, 2b, 2c as is observed experimentally.

Carbena-cyclohexadienones (3) react in a stereospecific manner too. The results for anthronylidene (3g) are not unambiguous (they point to a nonstereospecific addition), but the abstraction processes clearly call for a triplet state [33, 34]. The cycloaddition of cycloheptatrienylidene (4c) is stereospecific when dicyanoethylenes are the olefins used [61a, b] (the result of l. c. [36]) is obscured by base-induced isomerization of the fumarate and the maleate). Di- and tribenzo-cycloheptatrienylidene (*4a, b*), however, show a special behaviour. They add stereospecifically to olefins but in spite of this they give considerable amounts of abstraction products. Obviously the singlets *4a, b* add in a stereospecific manner, while the triplets *4a, b* lead only to the formation of abstraction products.

A new technique for a determination of spin-multiplicities is the CIDNP-method. From emission signals in the nmr-spectrum a singlet or triplet state is deduced in insertion or addition reactions. This technique has been applied recently to *3a* and *3g* [74]. CIDNP-measurements in cyclohexane or CCl_4 demonstrate that *3a* reacts as singlet whereas *3g* (di-*t*-butyl-derivative) involves a triplet intermediate. The latter experiment is in contrast with the results of stereo-specific addition of *3g* to *cis*- or *trans*-butene [68]. However the faster rate of addition versus insertion is thought to be responsible for this descrepancy [74].

If one wants to alter the S—T conversion rate, two experimental procedures are to hand:

a) *Dilution techniques* or

b) *sensitization experiments.*

With procedure a) an increase of nonstereospecificity from 34% to 77% for *2c* (*cis*-olefin) is achieved with hexafluorobenzene as diluting agent (see Table 11).

In contrast, *2a* showed almost no change in nonstereospecific addition when dilution experiments were carried out either with hexafluorobenzene or octafluorocyclobutane [66].

In cyclohexadienylidene *3g* however the stereospecifity in the addition to *cis*-butene is lost with increasing concentrations of hexafluorobenzene [68b]. The ratio of *cis*-adduct/*trans*-adduct drops from 18.0, 4.5 to 2.0 as the mole % of C_6F_6 is increased from 0, 74 to 90%. Solvents with heavy atoms such as methyl-iodide or phenyl-bromide favor also S → T conversions in the case of *3g* [68b].

A smaller increase is reported for *4c*; where nonstereospecific addition goes up from 2% to 6% (*cis*-olefin).

The dilution technique makes use of the different concentration dependence of the S—T conversion (α) and the carbene addition to the olefin (β). The decay of the metastable singlet state is unimolecular, while the stereospecific addition rate is first-order in olefin concentration [73]. The dilution technique has not yielded a common ratio in the experiments with *cis*- or *trans*-butene and *2c* (see Table 11). Extrapolation of the data to infinite dilution gives a product ratio of 0.16, suggesting that

even at high dilution the reaction proceeds to some extent via the singlet
2c.

b) While only the bimolecular reaction is influenced by the dilution
technique, the *sensitization procedure* can generate the triplet state
exclusively. This method has so far been applied only to di- and tetra-
phenylcyclopentadienylidene [45].

If a carbene — *e.g.* cyclopentadienylidene — is generated photochemically from the diazo precursor, intersystem crossing (S ⤳ T) can
occur either from the excited diazo singlet: *12** (S) ⤳ *12** (T), or from the
carbene singlet: *2* (S) ⤳ *2* (T) [51]. If a suitable sensitizing agent, *i.e.*
triplet energy-transfer agent, is employed, this problem is circumvented
since energy transfer gives the *12** triplet state which can only decompose
to *2* (T). All nonstereospecifity in the cyclo-addition with olefins — in the
presence of a sensitizer — is then due to the triplet state, provided there
exists no S—T equilibrium *2* (S) ⇄ *2* (T). If xanthone (triplet energy $E_T =$
74 kcal) is used with a suitable filter, excitation of *12f* occurs to a greater
extent (see Table 12) via energy transfer. The results of these experiments are shown in Table 12. The data collected in Table 12 show the
following:

Table 12. Triplet carbene reactions resulting from energy transfer

Carbene	Sensitizer (E_T in kcal/m)	*trans*	*cis*	Ref.
		Cyclopropane in (%) from		
		cis-olefin	*trans*-olefin	
	—	< 1[1]	< 1[1]	27)
	Xanthone (74)	27[1, 2]	$\ll$27[1, 2]	45)
2f				
2h	Benzophenone (68.5)	5—10[1, 2]	—	45)
H_2CN_2	—	0[3]	0[3]	65)
	Benzophenone (68.5)	34[3]	Trace[3]	76)
(ROOC)CH=N_2	—	0[3]	0[3]	77)
	Benzophenone	90[3]	14[3]	

[1]) *cis*- and *trans*-4-Methyl-pent-2-ene.
[2]) Light capture by the sensitizer was not complete (1:2) [45]).
[3]) *cis*- and *trans*-But-2-ene.

The amount of nonstereospecific addition of *2f* to *cis*- or *trans*-4-methyl-pent-2-ene is 27%, *i.e.* is due to the triplet *2f* produced by energy transfer. This seems to exclude any equilibrium between *2f* (S) $\rightleftarrows$ *2f* (T), since direct photolysis is stereospecific. Such an equilibrium has been postulated, however, for the temperature-dependent addition of diphenyl-carbene to *cis*-butene-(2) [75].

From competition experiments of *3g* with cyclohexene/pyridine–N-oxide mixtures, in contrast to *2f*, a singlet-triplet carbene equilibrium was concluded [78].

The following conclusions can be drawn from the multiplicity studies:

1. All cycloalkenecarbenes (but see p. 132) have a triplet ground state (ESR); despite this, they react at ambient temperature as singlets in a stereospecific manner.
2. Cycloalkenecarbenes not belonging to this group contain either fused aromatic rings or heavy-atom substituents. These groups can effect either a faster S $\rightsquigarrow$ T conversion or, less likely, a longer lifetime of the cycloalkenecarbene. Therefore S $\rightsquigarrow$ T conversion can occur more easily.
3. All nonstereospecifity results from two-step addition of triplet carbenes; a stereospecific two-step addition must be excluded.

139

4. The delocalized dipolar resonance formulae postulated on p. 104 for *1* and *4* are justified on the basis of the splitting of the carbene levels and the mild nucleophilicity of *4*. For the $4n+2$ carbene *2*, the carbene energy-level splitting is very small and no clear electrophilic character has been detected for *2*. Since *3* clearly behaves as an electrophil, a reasonable contribution from a delocalized resonance structure seems to be involved for *3*. Although no clear indication of electrophilic character has been observed for *2*, we believe that *2* is also best described by a delocalized resonance structure.

Die in dieser Arbeit zitierten Arbeiten wurden von der Deutschen Forschungsgemeinschaft und dem Fonds der Chemischen Industrie gefördert, wofür an dieser Stelle gedankt sei. Vielen Kollegen insbesondere Prof. W. M. Jones, University of Florida, Gainesville, USA danke ich für fruchtbare Diskussionen. Herrn Dr. W. E. Heyd, Herrn Dr. H. Kober und Herrn Dipl. Chem. W. Bujnoch danke ich für die Durchsicht und Frl. I. Halberstadt für das Schreiben des Manuskripts.

VI. References

[1] Kirmse, W.: Carbene, Carbenoide und Carbenanaloge. Weinheim/Bergstr.: Verlag Chemie 1969.

[2] Gilchrist, T. L., Reeves, C. W.: Carbenes, nitrenes and arynes. London: Th. Nelson and Sons 1969.

[3] For a recent summary see: Dürr, H. In: Methoden der Organ. Photochemie, Bd. IV/5, Houben Weyl, in press;
Reactions of o-quinone-diazides which undergo Wolff-rearrangement are equally excluded.

[4a] Gleiter, R., Hoffmann, R.: J. Am. Chem. Soc. *90*, 5457 (1968).

[4b] Keese, R., Krebs, E. P., Dürr, H.: unpublished results.

[5] Hoffmann, R., Zeiss, G. D., van Dine, G. W.: J. Am. Chem. Soc. *90*, 1485 (1968).

[6] Wasserman, E., Barash, L., Trozzolo, A. M., Murray, R. W., Yager, W. A.: J. Am. Chem. Soc. *86*, 2304 (1964); see also: Moser, R. E., Fritsch, J. M., Matthews, C. N.: Chem. Commun. *1967*, 770.

[7] Murray, R. W., Trozzolo, A. M., Wasserman, E.: J. Am. Chem. Soc. *84*, 4090 (1962).

[8] Wasserman, E., Trozzolo, A. M., Yager, W. A., Murray, R. W.: J. Chem. Phys. *40*, 2408 (1964).

[9] Murahashi, S., Moritani, I., Nagai, T.: Bull. Chem. Soc. Japan *40*, 1655 (1967).

[10] Moritani, I., Murahashi, S. I., Yamamoto, M. H. Y., Itoh, K., Mataga, N.: J. Am. Chem. Soc. *89*, 1259 (1967). — Brandon, R. W., Closs, G. L., Hutchinson, C. A.: J. Chem. Phys. *37*, 1878 (1962).

[11] Ebel, H. F.: Die Acidität der CH-Säuren. Stuttgart: G. Thieme Verlag 1969.

[12] Wasserman, E., Murray, R. W.: J. Am. Chem. Soc. *86*, 4203 (1964).

[13] Jones, M., Jr., Harrison, A. M., Rettig, K. R.: J. Am. Chem. Soc. *91*, 7462 (1969).

[14] Moriconi, E. J., Murray, J. J.: J. Org. Chem. *29*, 3577 (1964).

[15] Bernheim, R. A., Kempf, R. J., Granas, J. V., Skell, P. S.: J. Chem. Phys. *43*, 196 (1965).

16) Moritani, I., Murahashi, S. I., Yoshinaga, K., Ashitaka, H.: J. Chem. Soc. Japan *40*, 1506 (1967).

17) Closs, G. L., Hutchinson, C. A., Kohler, B. E.: J. Chem. Phys. *44*, 413 (1966).

18) Trozzolo, A. M., Gibbons, A. W.: J. Am. Chem. Soc. *89*, 239 (1967).

19) Moritani, I., Murahashi, S. I., Ashitaka, H., Kimura, K., Tsubomera, H.: J. Am. Chem. Soc. *90*, 5918 (1968).

20) Moritani, I., Murahashi, S., Nishino, M., Kimura, K.: Tetrahedron Letters *1966*, 373.

21) Yamamoto, Y., Moritani, I., Maeda, Y., Murahashi, S.: Tetrahedron *26*, 251 (1970).

22) Jones, W. M., Stowe, M. E.: Tetrahedron Letters *1964*, 3459.

23) Jones, W. M., Denham, J. M.: J. Am. Chem. Soc. *86*, 944 (1964).

24) McGregor, S. D., Jones, W. M.: J. Am. Chem. Soc. *90*, 123 (1968).

25) Jones, W. M., Stowe, M. E., Wells, E. E., Jr., Lester, E. W.: J. Am. Chem. Soc. *90*, 1849 (1968).

26) Moss, R. A.: J. Org. Chem. *31*, 3296 (1966).

27) Dürr, H., Scheppers, G.: Chem. Ber. *100*, 3236 (1967).

28) McBee, E. T., Bosoms, J. A., Morton, C. J.: J. Org. Chem. *31*, 768 (1966), and McBee, E. T., Sienkowski, K. J.: J. Org. Chem. *38*, 1340 (1973).

29) Doering, W. v. E., Jones, M., Jr.: Tetrahedron Letters *1963*, 791.

30) Lloyd, D.: Carbocyclic nonbenzenoid aromatic compounds, p. 55. Amsterdam: Elsevier Publ. Comp. 1966.

31) Süs, O., Möller, K., Heiss, H.: Liebigs Ann. Chem. *598*, 123 (1956).

32) Dewar, M. J. S., Narayanaswami, K.: J. Am. Chem. Soc. *86*, 2422 (1964); see also: Baldwin, J. E., Smith, R. A.: J. Am. Chem. Soc. *89*, 1886 (1967).

33) Cauquis, G., Reverdy, G.: Tetrahedron Letters *1967*, 1493.

34) Fleming, J. C., Shechter, H.: J. Org. Chem. *34*, 3962 (1969).

35) Jones, W. M., Ennis, C. L.: J. Am. Chem. Soc. *89*, 3069 (1967).

36) Mukai, T., Nakazawa, T., Isobe, T.: Tetrahedron Letters *1968*, 565.

37) Murahashi, S. I., Moritani, I., Nishino, M.: J. Am. Chem. Soc. *89*, 1257 (1967).

38) Joines, R. C., Turner, A. B., Jones, W. M.: J. Am. Chem. Soc. *91*, 7754 (1969); Schissel, P. O., Kent, M. E., Mc Adoo, D. J., Hedaya, H.: J. Am. Chem. Soc. *92*, 2147 (1970). — Mitsuhashi, T., Jones, W. M.: J. Am. Chem. Soc. *94*, 677 (1972). — Krajca, K. E., Mitsuhashi, T., Jones, W. M.: J. Am. Chem. Soc. *94*, 3661 (1972). — Vander Stouw, G. G., Kraska, A. R., Shechter, H.: J. Am. Chem. Soc. *94*, 1655 (1972). — Jones, W. M., *et al.*: J. Am. Chem. Soc. *95*, 826 (1973).

39) LaBar, R. A., Jones, W. M.: J. Am. Chem. Soc., *95*, 2359 (1973). — Gebert, P. H., King, R. W., Jones, W. M.: J. Am. Chem. Soc., *95*, 2357 (1973)

40) Schöherr, H. J., Wanzlick, H. W.: Chem. Ber. *103*, 1037 (1970).

41) Wanzlick, H. W., Kleiner, H. J.: Angew. Chem. *75*, 1024 (1963).

42) Quast, H., Frankenfeld, E.: Angew. Chem. *77*, 680 (1965).

43) Kirmse, W., Horner, L., Hoffmann, H.: Liebigs Ann. Chem. *614*, 19 (1958).

44) Dürr, H., Schrader, L.: Chem. Ber. *102*, 2026 (1969).

45) Dürr, H., Bujnoch, W.: Tetrahedron Letters *1973* and unpublished material.

46) Doering, W. v. E., Buttery, R. G., Laughlin, R. G., Chaudhuri, N.: J. Am. Chem. Soc. *78*, 3224 (1965).

47) Doering, W. v. E., Knox, L. H.: J. Am. Chem. Soc. *83*, 1989 (1961).

48) Dietrich, H., Griffin, G. W., Petterson, R. C.: Tetrahedron Letters *1968*, 153.

49) Ciganek, E.: J. Am. Chem. Soc. *88*, 1979 (1966).

50) Dürr, H., Herrmann, W.: unpublished results.

51) Frey, H. M.: Proc. Roy. Acad. *250*, 409 (1959).

52) Frey, H. M.: J. Am. Chem. Soc. *80*, 5005 (1958).

H. Dürr:

53) Doering, W. v. E., Prinzbach, H.: Tetrahedron 6, 24 (1959).
54) Dürr, H., Sergio, R., Scheppers, G.: Liebigs Ann. Chem. 740, 63 (1970).
55) Dürr, H., Scheppers, G.: Angew. Chem. 80, 359 (1968); Liebigs Ann. Chem. 734, 141 (1970).
56) Moss, R. A.: In: Selective organic transformations, Vol. I, p. 35 (ed. by B. S. Thyagarajan). Wiley-Interscience 1970.
57) Closs, G. L., Coyle, J. J.: J. Am. Chem. Soc. 87, 4270 (1965).
58) Levin, R. H., Jones, M., Jr.: private communication.
59) Closs, G. L., Moss, R. A.: J. Am. Chem. Soc. 86, 4042 (1964).
60) Moss, R. A., Gerstl, R.: Tetrahedron Letters 1967, 4905; Tetrahedron 22, 2637 (1966).
61a) Jones, W. M., Ennis, C. L.: J. Am. Chem. Soc. 91, 6391 (1969).
61b) Jones, W. M., Hamon, B. N., Joines, R. C., Ennis, C. L.: Tetrahedron Letters 1969, 3909.
62) Christensen, L. W., Waali, E. E., Jones, W. M.: J. Am. Chem. Soc. 94, 2118 (1972); for the method used see also: Seyferth, D., Mui, Y. P., Damrauer, R.: J. Am. Chem. Soc. 90, 6182 (1968). — Sadler, I. H.: J. Chem. Soc. (B) 1969, 1024.
63) See also: Kunitake, T., Price, C. C.: J. Am. Chem. Soc. 85, 1318 (1963).
64) Quast, H., Hünig, S.: Chem. Ber. 99, 2017 (1966).
65) Skell, P. S., Woodworth, R. C.: J. Am. Chem. Soc. 78, 4496, 6427 (1956); 81, 3383 (1959).
66) Moss, R. A. Przybyla, J. R.: J. Org. Chem. 33, 3816 (1968).
67) Jones, M., Jr., Rettig, K. R.: J. Am. Chem. Soc. 87, 4013 (1965).
68a) Koser, G. F., Pirkle, W. H.: J. Org. Chem. 32, 1992 (1967).
68b) Pirkle, W. H., Koser, G. F.: Tetrahedron Letters 1968, 3959.
69) Cauquis, G., Reverdy, G.: Tetrahedron Letters 1968, 1085.
70) Gutsche, C. D., Bachman, G. L., Coffey, R. S.: Tetrahedron Letters 18, 617 (1962).
71) Closs, G. L., Closs, L. E.: Angew. Chem. 74, 431 (1962).
72) Ciganek, E.: J. Am. Chem. Soc. 88, 1979 (1966).
73) Heilbronner, E., Bock, H.: Das HMO-Modell und seine Anwendung. Weinheim/ Bergstr.: Verlag Chemie 1970.
74) Kaplan, M. L., Roth, H. D.: Chem. Commun. 1972, 970.
75) Closs, G. L.: Topics in stereochemistry, Vol. 3, p. 193. New York: Interscience Publishers 1968.
76) Kopecky, K. R., Hammond, G. S., Leermakers, P. A.: J. Am. Chem. Soc. 83, 2397 (1961); 84, 1015 (1962).
77) Jones, M., Jr., Ando, W., Kulczycki, A.: Tetrahedron Letters 1967, 1391.
78) Cauquis, G., Reverdy, G.: Tetrahedron Letters 1971, 3771.

Received November 30, 1972

HMO
Hückel Molecular Orbitals

From the reviews:

**E. Heilbronner
and P. A. Straub**

With 816 pages
DIN A 4
Loose Leaf. 1966
DM 92,—

"In 1961, when Streitwieser wrote Molecular Orbital Theory for Organic Chemists, he drew attention to the very rapid recent growth of interest in this field—seventy papers in the forties, 600 in the fifties, and a corresponding increase in the sixties. These π-electron molecular orbitals are usually represented as linear combinations of atomic orbitals (LCAO) with certain other approximations as introduced by Hückel. The enormous use of these Hückel MOs has now led to no less than three fullscale publications of tables of the relevant coefficients. The present volume, prepared by Prof. Heilbronner and Dr. Straub, is the latest attempt to provide the coefficients in the MOs and certain other dependent quantities in such a form as to be helpful to chemists who have no desire to make these calculations for themselves." (Nature)

Springer-Verlag
Berlin · Heidelberg · New York
München · London · Paris · Sydney · Tokio · Wien

Topics in
Current Chemistry
Fortschritte der
chemischen Forschung

Author Index
and
Subject Index

Volumes 31-40

Springer-Verlag Berlin Heidelberg New York

Author Index Volumes 31-40

3

Subject Index Volumes 31-40

The volume numbers are printed in italics